Fisheries Biotechnology and Bioinformatics

C Judith Betsy • C. Siva

Fisheries Biotechnology and Bioinformatics

Springer

C Judith Betsy
TNJFU - Fisheries College and Research Institute
Tuticorin, Tamil Nadu, India

C. Siva
ICAR - Directorate of Coldwater Fisheries Research
Nanital, Uttarakhand, India

ISBN 978-981-99-6993-7 ISBN 978-981-99-6991-3 (eBook)
https://doi.org/10.1007/978-981-99-6991-3

This Springer imprint is published by the registered company Springer Nature Singapore Pte Ltd.
The registered company address is: 152 Beach Road, #21-01/04 Gateway East, Singapore 189721, Singapore

Paper in this product is recyclable.

Preface

Biotechnology is a multidisciplinary subject which includes biology, microbiology, biochemistry, molecular biology, genetics, chemistry, and process engineering. It all started with beer fermentation, which opened up new avenues in biotechnology. Fermenters have been developed to produce penicillin to save lives. Biotechnology has revolutionized different fields, including agriculture, veterinary, and fisheries. This technology helps change organisms' genetic makeup to produce new traits/characters. The biotechnological developments made in agriculture and veterinary are discussed with special emphasis on fisheries, enlightening readers about the journey the technology has made in different fields.

Nucleic acids are the basis of biotechnology. We have tried to describe every single component involved in nucleic acids. A detailed description of the structure and function of DNA and RNA and their structural and functional differences is provided. Possible information on the conformations and different types of DNA and RNA is also provided. The information about genes and genetic codes is dealt in detail in which the activities, structure, and different types of genes and their functions, codons, and wobble pairing are explained. This will give an entry-level understanding of the process behind cell mechanisms. The mutation is an essential aspect that must be understood to work with many different tools involved in biotechnology. Hence, the effect of mutation and its types at both gene and chromosome levels are explained with utmost care.

Studies on gene expression in many organisms have become inevitable. Therefore, understanding the mechanism of gene expression in both prokaryotes and eukaryotes is necessary. Although challenging, effort has been made to make these topics understandable with as many diagrammatic representations as possible. In the case of prokaryotes, the details on formation, components, structure of operons, and different types of operons like lac and trp operons are provided. In eukaryotes, the complete process of replication, transcription, and translation of DNA is provided. Understanding epigenetics is essential to know the interaction of genes with their environment. Keeping this in mind, the basics of DNA methylation and histone modifications are provided to impart knowledge. The role of RNA in gene regulation and different types of antisense RNA is also reported.

The primary technologies, such as recombinant DNA technology, transgenic fish production, animal cell culture, hybridoma technology, and cryopreservation technology, are narrated. Similarly, various molecular techniques, including PCR, ELISA, blotting techniques, DNA fingerprinting, and DNA and protein sequencing, are discussed. Information on different molecular markers like RFLP, RAPD, AFLP, microsatellites, ESTs, and SNPs is also provided. Industrial biotechnological tools such as biosensors, bioremediation, bioprospecting, and bioprocess engineering are individually dealt with.

Bioinformatics is an interdisciplinary field mainly involving molecular biology and genetics, computer science, mathematics, and statistics, where data-intensive, large-scale biological problems are addressed from a computational point of view. It stores, distributes, and analyzes biological data to support many research areas. Different databases and formats are available for storing and searching the data; the tools and techniques used for analyzing the data are emphasized. The book also discusses genomics and its application in fisheries.

Given its coverage and depth, Fisheries Biotechnology and Bioinformatics is a classical textbook. It follows the arrangements and sequence of typical textbooks to allow easier adoption by faculty and for ease of use by students. The targeted users of this book are both undergraduate and postgraduate students. We have used as many figures as possible from various published research papers, and appropriate credit has been given to the figures. Although we have tried our best to make the book error free, we apologize if any error is found. Your constructive comments will help us improve the book.

Tamil Nadu, India — Judith Betsy
Nanital, India — C. Siva

Contents

1 Biotechnology in Fisheries 1
1.1 Biotechnology 1
1.2 Concepts and Terminologies in Biotechnology 1
1.3 Fields of Biotechnology 2
1.3.1 Agricultural Biotechnology 2
1.3.2 Animal Biotechnology 3
1.3.3 Industrial Biotechnology 3
1.3.4 Environmental Biotechnology 3
1.3.5 Fisheries Biotechnology 3
1.4 Significant Developments in the Field of Biotechnology 4
1.5 Significant Developments in Fisheries Biotechnology 5
Reference 6

2 Nucleic Acids 7
2.1 Components of Nucleic Acid 7
2.1.1 Nitrogenous Bases 7
2.1.2 Pentose Sugar 7
2.1.3 Phosphoric Acid 8
2.2 Nucleoside and Nucleotide 8
2.3 Polynucleotide or Nucleic Acid Strand 8
2.4 DNA as Genetic Material 9
2.4.1 DNA Structure 10
2.4.2 Different Forms of DNA 13
2.5 Ribonucleic Acid (RNA) 15
2.5.1 Chemical Structure of RNA 16
2.5.2 Types of RNA 16
2.6 Difference Between DNA and RNA 17
References 18

3 Genes and Genetic Codes 19
3.1 Genes 19
3.1.1 Activities of Genes 19
3.1.2 Structure of Gene 20

3.1.3 Types of Genes Based on Their Function and Position 20
3.2 Genetic Code 21
3.2.1 Discovery of the Genetic Code 22
3.2.2 Codons and Amino Acids 22
3.2.3 Properties of the Genetic Code 22
3.2.4 The Wobble Hypothesis 23
References 25

4 **Mutations and Their Implications** 27
4.1 Effects of Mutations 27
4.2 Types of Mutations 28
4.2.1 Gene Mutations 28
4.2.2 Chromosomal Mutations (Aberrations) 30

5 **Gene Expression in Prokaryotes** 35
5.1 Formation of Operon 35
5.2 Components of Operon 36
5.3 Structure of Operon 36
5.4 Functions of Operon 36
5.5 Lac Operon 37
5.6 Trp Operon 38
References 40

6 **Gene Expression in Eukaryotes** 41
6.1 DNA Replication, Transcription, and Translation 41
6.1.1 Replication 41
6.1.2 Transcription 44
6.1.3 Translation 46
References 48

7 **Epigenetics** 49
7.1 Covalent Modification of DNA: Cytosine Methylation 49
7.2 Methylation Regulates Transcription Through Multiple Mechanisms 50
7.3 DNA Methylation in Fishes 50
7.4 Histone Modifications 51
7.4.1 Acetylation 52
7.4.2 Methylation 52
7.4.3 Ubiquitination 53
7.4.4 Phosphorylation 54
7.4.5 Histone Post-translational Modifications in Fish 54
References 54

8 **RNA in Gene Regulation** 57
8.1 Antisense RNA 57
8.1.1 Small Interfering RNA 58

8.1.2 Micro RNAs . . . 58
8.1.3 lncRNAs . . . 59
8.1.4 piRNAs . . . 59
8.1.5 Ribozymes . . . 60
References . . . 61

9 Recombinant DNA Technology . . . 63
9.1 Enzymes . . . 63
9.1.1 Restriction Enzymes . . . 63
9.1.2 DNA Polymerase . . . 64
9.1.3 Taq DNA Polymerase . . . 64
9.1.4 Reverse Transcriptase . . . 64
9.1.5 DNA Ligase . . . 64
9.2 Vectors . . . 65
9.2.1 Plasmids . . . 65
9.2.2 Types of Cloning Vectors . . . 66
9.3 Host Organism . . . 71
9.4 Steps Involved . . . 71
9.4.1 Isolation of the Genetic Material . . . 73
9.4.2 Cutting of DNA at Specific Locations . . . 73
9.4.3 Isolation of Desired DNA Fragment . . . 73
9.4.4 Amplification of Gene of Interest Using PCR . . . 73
9.4.5 Ligation of DNA Fragment into a Vector . . . 73
9.4.6 Insertion of Recombinant DNA into the Host Cell . . . 73
9.4.7 Culturing the Foreign Gene Product . . . 74
9.4.8 Cloning Foreign DNA into Plasmid DNA . . . 74
References . . . 75

10 Transgenic Fish Production . . . 77
10.1 Selection of Fish Species . . . 77
10.2 Preparation of Transgene Constructs . . . 78
10.3 Isolation of the Gene of Interest . . . 79
10.3.1 Genes of Interest . . . 79
10.4 Cloning the Gene of Interest . . . 80
10.5 Gene Transfer . . . 80
10.5.1 Microinjection . . . 80
10.5.2 Electroporation . . . 81
10.5.3 Transgenic Mosaicism . . . 81
10.6 Identification of Transgenic Fish . . . 82
10.7 Expression of Transgenes . . . 82
10.8 Inheritance of Transgenes . . . 82
10.9 Advantages . . . 82
10.10 Risks . . . 83

11 Animal Cell Culture . . . 85
11.1 Types of Animal Cell Culture . . . 85

11.1.1 Based on the Number of Cell Division 85
11.1.2 Based on the Life Span of Culture 86
11.2 Cell Strain 87
11.2.1 Methods 87
11.3 Applications of Cell Line 88
11.3.1 Vaccines Production 88
11.3.2 Virus Cultivation and Study 88
11.3.3 Cellular and Molecular Biology 89
11.3.4 Cancer Research 89
11.3.5 Gene Therapy 89
11.3.6 Immunological Studies 89
11.3.7 Others 89

12 Hybridoma Technology 91
12.1 Monoclonal Antibodies (MAbs) 91
12.2 Production of MAb 92
12.2.1 Steps in Production of MAb 92
12.3 Application of Monoclonal Antibodies (MAbs) in Aquaculture 94

13 Cryopreservation Technology in Fishes 97
13.1 Diluents 98
13.1.1 Extenders 98
13.1.2 Cryoprotectants 100
13.2 Dilution 103
13.3 Process of Cryopreservation 104
13.3.1 Selection of Brooders 104
13.3.2 Milt Collection 104
13.3.3 Diluent Preparation 107
13.3.4 Equilibration of Milt 107
13.3.5 Packing of Milt 107
13.3.6 Pre-freezing 108
13.3.7 Freezing and Storage 109
13.3.8 Thawing 109
13.4 Cryopreservation in Aquaculture 110
References 111

14 Molecular Techniques 115
14.1 Polymerase Chain Reaction (PCR) 115
14.1.1 Requirements of a PCR Reaction 115
14.1.2 PCR Cycle 115
14.1.3 Steps in PCR 117
14.1.4 Types of PCR 118
14.1.5 Uses of PCR 118
14.2 Enzyme-Linked Immunosorbent Assay (ELISA) 119
14.2.1 ELISA Principle 119

14.2.2 Requirements for ELISA 119
14.2.3 Stages in ELISA 120
14.2.4 Methods in ELISA 120
14.2.5 Types of ELISA Assay 121
14.3 Blotting Techniques 123
14.3.1 Southern Blotting 124
14.3.2 Northern Blotting 125
14.3.3 Western Blotting 126
14.4 DNA Fingerprinting 128
14.4.1 Procedure 128
14.4.2 Application in Fisheries 130
14.5 DNA Sequencing 130
14.5.1 Maxam–Gilbert Sequencing 131
14.5.2 Chain Termination Method 131
14.5.3 Dye-Terminator Sequencing 132
14.5.4 Pyrosequencing 132
14.5.5 High-Throughput Sequencing 134
14.5.6 Applications of DNA Sequencing Technologies (Mardis 2017) 137
14.6 Protein Sequencing 137
14.6.1 Protein Sequencing by Edman Degradation 137
14.6.2 Mass Spectrometry (MS) 138
14.6.3 Applications of Protein Sequencing 139
References 139

15 Molecular Markers 141
15.1 Advantages of Molecular Markers 141
15.2 Properties of Ideal Genetic Markers 141
15.3 Types of Markers 142
15.3.1 Non-PCR-Based Markers 142
15.3.2 PCR-Based Markers 144
15.3.3 Sequence-Based Markers 149
References 150

16 Biosensor 153
16.1 Immobilization of Biological Material 153
16.2 Components of a Biosensor 154
16.3 Construction of Biosensor 154
16.3.1 Pre-requisites for Developing Biosensor 154
16.3.2 Characteristics of Good Biosensor 155
16.4 Types of Biosensors 155
16.4.1 Calorimetric Biosensors 155
16.4.2 Potentiometric Biosensors 155
16.4.3 Amperometric Biosensors 156
16.4.4 Optical Biosensors 156

16.4.5 Acoustic Wave Biosensors 156
16.5 Applications 156
16.6 Applications in Fisheries 157
16.6.1 Biosensors in Food Analysis 157
16.6.2 Biosensors in Aquatic Environment Assays 157
16.6.3 Aequorin 157
16.6.4 Green Fluorescent Protein (GFP) 158
16.6.5 Biosensors in the Detection of Fish Toxins 158
16.6.6 Lab-on-a-Fish from PNNL 158
16.6.7 BIOLAN for Sulfite Testing 159
References 159

17 Bioremediation 161
17.1 Principle of Bioremediation 161
17.1.1 Factors Determining the Success of Bioremediation . . 161
17.2 Bioremediation in Fisheries 162
17.2.1 Bioremediation Using Algal Mat 162
17.2.2 Bioremediation Using Bagasse 162
17.2.3 Bioremediation of Wastewater Using Aquatic Weeds 163
17.2.4 Bioremediation of Wastewater Using Earthworms . . . 163
17.2.5 Bioremediation of Spilled Oil 163
17.2.6 Bioremediation of Organophosphorus Pesticides 163
17.2.7 Other Bioremediation 164
17.3 Advantages of Bioremediation 164
17.4 Disadvantages of Bioremediation 164
References 165

18 Bioprospecting in Fisheries 167
18.1 Phases Involved 168
18.2 Bioprospecting Marine Biological Resources 168
18.3 Regulation of Bioprospecting 171
18.3.1 Bonn Guidelines (Sirakaya 2019) 171
18.3.2 United Nations Convention on the Law of the Seas (UNCLOS) (Guilfoyle 2019) 172
18.3.3 Global Ocean Commission (Global Ocean Commission 2014) 172
18.3.4 International Sea Bed Authority (Jaeckel 2020) 172
18.3.5 Valencia Declaration (Laladhas et al. 2016) 172
18.3.6 European Micro B3 (Chege kamau et al. 2015) 172
18.3.7 European Science Foundation 173
18.3.8 Nagoya Protocol (Martins et al. 2020) 173
18.4 Biopiracy 173
References 174

19 Bioprocess Engineering in Fisheries . . . 175
19.1 Stages Involved . . . 175
19.1.1 Upstream Processing . . . 175
19.1.2 Reactor Engineering . . . 178
19.1.3 Downstream Processing . . . 182
19.2 Role of Bioprocess Engineering in Fisheries . . . 185
References . . . 186

20 Bioinformatics: Databases and Formats . . . 187
20.1 Databases . . . 187
20.2 Classifications of Databases . . . 188
20.2.1 Primary Database . . . 188
20.2.2 Secondary Database . . . 188
20.2.3 Biological Databases . . . 188
20.3 Tools for Sequence Analysis . . . 191
20.3.1 BLAST (Basic Local Alignment Search Tool) . . . 191
20.3.2 CLUSTALW . . . 192
20.3.3 FASTA . . . 192

21 Sequence Retrieval, Analysis, and Manipulation . . . 193
21.1 Sequence Retrieval System . . . 194
21.1.1 Types of Sequence Retrieval Databases . . . 194
21.1.2 Sequence Retrieval in NCBI . . . 195
21.2 Sequence Analysis . . . 195
21.2.1 Output of Sequence Analysis . . . 196
21.2.2 Parameters for Which Sequence Analysis Is Applied . . . 196
21.3 Sequence Manipulation Suite . . . 198
21.3.1 DNA Entry . . . 198
21.3.2 DNA Manipulation . . . 198
21.3.3 DNA Figures . . . 198
21.3.4 DNA Analysis . . . 199
21.3.5 Protein Entry . . . 199
21.3.6 Protein Manipulation . . . 199
21.3.7 Protein Figures . . . 199
21.3.8 Protein Analysis . . . 199

22 Bioinformatics Tools and Techniques . . . 205
22.1 Primer Design . . . 205
22.1.1 Before Starting . . . 205
22.1.2 Characteristics of Good Primer . . . 205
22.1.3 Good Primer . . . 207
22.2 Restriction Mapping . . . 208
22.2.1 Method . . . 208
22.2.2 Example . . . 209
22.3 Orf Finder . . . 210

22.3.1 Reading Frame 210
22.3.2 Open Reading Frames 210
22.3.3 ORF Finding Tools 211
22.4 EMBOSS 212
22.4.1 Advantages of EMBOSS 212
22.4.2 Applications of EMBOSS 213
22.5 Molecular Visualization 213
22.5.1 Tools for Molecular Visualization 213
22.6 Benefits of Bioinformatics 214
22.7 Future Scope 214

23 Genomics and Its Application in Fisheries 217
23.1 Application in Aquaculture and Fisheries 217
23.2 Potential Genomic Technologies for Aquaculture and Fisheries 218
23.3 Major Aquaculture Genome Projects 218
23.4 Fish Genome Sequencing 219
References 220

About the Authors

Judith Betsy is an Assistant Professor at the Department of Aquaculture, Fisheries College and Research Institute of Tamil Nadu Dr. J. Jayalalithaa Fisheries University, Tuticorin, India. Holding a doctorate in aquaculture, she is a recipient of the Maulana Azad National Fellowship from the University Grants Commission, New Delhi, India. Having worked in the area of fish gamete cryopreservation and spermatology for the past 10 years, she has been honored with 10 awards including gold medals for her outstanding academic performance during her graduate studies and has published over 30 research papers in peer-reviewed journals around the world, more than 85 technical articles, 6 book chapters, and 6 books.

C. Siva is a Scientist at the Indian Council of Agricultural Research—Directorate of Coldwater Fisheries Research, Bhimtal, India. He has more than 8 years of experience in the field of fish genetics and breeding. He has operated around ten institutionally and externally funded research projects in various capacities. He has been involved in developing complete mitogenomes, transcriptome assemblies, and molecular markers for important cold-water fishes of India. In addition, he is also involved in studies related to the identification of different fish genetic stocks in India and mining genes and alleles related to economic traits and sex-related functions in fish. Through all these activities, he has several research publications to his credit. At present, he is working in the area of comparative transcriptomics to study the underlying genetic mechanisms of sex determination and differentiation in important cold-water fish species of India.

Biotechnology in Fisheries 1

The history of fermentation dates back to 10,000 BC. People from various parts of the world have been using fermented products. It was Louis Pasteur who scientifically reported fermentation process. Their understanding on fermentation lead to fermenting beer and opened up a new field named Zymo technology. Zymo technology is the field of study about the process of fermentation in different products. It is from this technology, biotechnology developed.

1.1 Biotechnology

The term biotechnology was coined by Karl Ereky in 1989. He described this term to refer a process or technology that biologically converts raw materials into products. Biotechnology is the application of scientific and engineering principles (microbiology, genetics, biochemistry, etc.) to the processing of materials by biological agents (microorganisms, enzymes, plant and animal cells) to provide goods and service. The major timeline of development in biotechnology are as follows:

- 1920—Leads City Council, U.K. established the Institute of Biotechnology
- 1970s—Biotechnology emerged as a new discipline
- 1978—European Federation of Biotechnology was established
- 1982—Government of India established the National Biotechnology Board and in 1986, it became a separate department, Department of Biotechnology in the Ministry of Science and Technology

1.2 Concepts and Terminologies in Biotechnology

- *Genetic engineering* is the modification or manipulation of animals genome by employing DNA technology and is also called genetic modification.

C. J. Betsy, C. Siva, *Fisheries Biotechnology and Bioinformatics*, https://doi.org/10.1007/978-981-99-6991-3_1

- *Molecular biology* is the branch of biology that deals with the formation, composition, structure, and function of macromolecules.
- *Genomics* is the study of the genome which is the complete set of DNA of an organism in which the genomes are mapped, sequenced, and analyzed.
- *Transcriptomics* is the study of transcriptome which is the set of all RNA transcripts, including coding and non-coding regions in cells to investigate gene expressions.
- *Proteomics* is the study of proteomes which is a set of proteins produced in an organism. Williams and Hochstrasser first used proteomic in 1994.
- *Metabolomics* is the study of small molecules called metabolites and its comprehensive analysis in a biological sample. These small molecules and their interactions within a biological system are known as the metabolome.
- *Functional genomics* integrates molecular biology and cell biology and deals with the whole structure, function, and regulation of a gene in contrast to the gene-by-gene approach of classical molecular biology technique. It aims to relate the phenotype and genotype on genome level and includes processes such as transcription, translation, protein–protein interaction, and epigenetic regulation (Kaushik et al. 2019).
- *Metagenomics* is the study of the structure and function of entire nucleotide sequences isolated and analyzed from all the organisms especially microbes.

1.3 Fields of Biotechnology

1.3.1 Agricultural Biotechnology

Agricultural biotechnology started with selecting the most suitable plants and animals for breeding. This was followed by the identification of principles of inheritance by Gregor Mendel which led to major advances in plant breeding by recognizing desirable traits and incorporating them into future generations. Through mutation, some of these traits can spontaneously occur. However, in nature, the rate of mutation is very slow and unreliable to produce all plants. Later it was discovered that exposing plants to X-rays and chemicals could increase the rate of genetic variation which in turn increases the characteristic pool from which breeders can select from. Some of the plant varieties produced using mutation breeding include wheat, barley, rice, potatoes, soybeans, and onions.

With the help of biotechnology, the crop yields were increased and the land use was reduced. Malnutrition in underdeveloped countries is also being combated with biotechnology. Golden rice, a rice variety which incorporates Vitamin A is developed by the International Rice Research Institute (IRRI) and its national research partners. Various organizations have approved it to be safe with added benefits of beta carotene, and the Philippines is the first country to approve golden rice for commercial use. Similarly, many genetic manipulation works are aimed at making crops more productive by reducing their dependence on pesticides, fertilizers, and irrigation, or by increasing their disease resistance.

1.3.2 Animal Biotechnology

The animal biotechnology has the following applications:

- Development of vaccines to protect animals from disease
- Production of superior calves through superovulation and embryo transfer technology
- Production of several calves from one embryo using embryo splitting or cloning
- Increase of animal growth rate
- Rapid disease detection by molecular immunological techniques
- Monoclonal antibody production
- Recombinant vaccine production
- Transgenic technology in animal production

1.3.3 Industrial Biotechnology

Industrial biotechnology aims to improve the efficiency and reduce the environmental impacts of industrial processes like textile, paper, and pulp, and chemical manufacturing. By applying biotechnology, companies develop enzymes in sufficient quantities to synthesize chemicals. Production of improved variety of cotton that is warmer, stronger, has improved dye uptake and retention, enhanced absorbency, and wrinkle- and shrink-resistance, production of biofuel from algae and use of corn in petroleum to produce chemicals are some of the applications of industrial biotechnology.

1.3.4 Environmental Biotechnology

Bioremediation is a major biotechnological application employed in clearing the environment. Some of the application of environmental biotechnology includes using enzyme bioreactors to treat industrial waste and food waste, converting waste to biofuel to run generators, inducing microbes to produce enzymes that can convert plant and vegetable materials into building blocks for biodegradable plastics, and finally pollution detecting and monitoring.

1.3.5 Fisheries Biotechnology

Fisheries biotechnology can be broadly classified into aquaculture biotechnology, marine biotechnology, algal biotechnology, and processing biotechnology. Biotechnology has played a major role in the areas of induction and control of maturation and spawning, chromosome manipulation, sex reversal in hermaphrodite fishes, production of triploid, tetraploid, and transgenic fishes. Some of the traits commonly used for producing improved varieties of fish are growth, disease, and cold tolerance.

The first genetically engineered fish to be sold commercially is the fluorescent GloFish®, a zebra fish modified to glow red, which came into the US market in 2004. Other areas include disease diagnosis (molecular and immunodiagnostic kits), hybridoma technology, and production of probiotics, vaccines and immunostimulants, cryopreservation of fish gametes, cell and tissue culture, and bioprospecting.

1.4 Significant Developments in the Field of Biotechnology

- 6000 BC—Bread making involving yeast fermentation
- 1857 AD—Pasteur proves that yeasts are living cells that cause alcohol fermentation
- 1928—Alexander Fleming discovers penicillin from *Penicillium notatum*
- 1953—DNA structure and function elucidated
- 1970—Hamilton Smith and colleagues determined that restriction enzymes could cleave DNA molecules at precise and predictable locations and the enzymes were able to recognize specific nucleotide sequences
- 1970—Smith et al. report restriction endonuclease from *Haemophilus influenzae* that recognizes specific DNA target sequences
- 1972—Walter Fiers and his team at the Laboratory of Molecular Biology of the University of Ghent, Belgium were the first to determine the sequence of the gene for bacteriophage MS2 coat protein.
- 1973—Herb Boyer used restriction enzymes to produce a DNA molecule with genetic material from two different sources and named it as recombinant DNA technology
- 1976—Walter Fiers and his team determined the complete nucleotide sequence of bacteriophage MS2-RNA
- 1975—Kohler and Milstein reported monoclonal antibodies
- 1977—DNA is sequenced for the first time by Fred Sanger, Walter Gilbert, and Allan Maxam working independently. Sanger's lab sequenced the entire genome of bacteriophage Φ-X174
- 1979—Paulien Hogeweg coined the term bioinformatics for the study of informatic processes in biotic systems
- 1980—Gordon et al. revolutionized the procedure for producing transgenic animals based on the microinjection of cloned DNA into the pronucleus of fertilized eggs at the one-cell stage
- 1982—Palmiter and his team produced first transgenic mouse by transferring metallothionein-I gene from rat to mouse
- 1983—Kary B. Mullis discovered the polymerase chain reaction
- 1984—Production of transgenic pig, rabbit, and sheep by microinjection of foreign DNA into egg nuclei
- 1989—The human gene that encodes the CFTR protein was sequenced by Francis Collins and Lap-Chee Tsui. Defects in this gene cause cystic fibrosis

- 1995—The genome of *Haemophilus influenzae* is the first genome of a free living organism to be sequenced
- 1996—*Saccharomyces cerevisiae* is the first eukaryote genome sequence to be released
- 1998—The first genome sequence for a multicellular eukaryote, *Caenorhabditis elegans*, is released
- 2001—First draft sequences of the human genome are released simultaneously by the Human Genome Project and Celera Genomics
- 2003—Successful completion of Human Genome Project with 99% of the genome sequenced to a 99.99% accuracy

1.5 Significant Developments in Fisheries Biotechnology

- 1973—Tong and his team injected mRNA and rRNA from mature eggs of crucian carp and common carp into newly fertilized crucian carp eggs, in order to induce character variation in goldfish
- 1973—Tong and Niu, transplanted nuclei between gold fish (*Carassius auratus*) and *Rhodeus sinensis* for the purpose of studying the developmental variations between the integrated nuclei and the pure heterologous nuclei, and the effects of cytoplasm on the nucleus
- 1982—Induced the first viable tetraploid fish, rainbow trout
- 1984—Maclean and Talwar reported microinjection of cloned DNA into rainbow trout (*Oncorhynchus mykiss*) eggs.
- 1985—Zhu produced the first transgenic fish, goldfish
- 1986—Fletcher and his team showed that AFP injection to seawater-acclimatized rainbow trout lowered the freezing point of the whole fish in proportion to the circulating anti-freeze protein concentration.
- 1986—Chen and his team transplanted cell nuclei from a grass carp blastula cell line into unfertilized, enucleated eggs of crucian carp, thus creating the first "test-tube fish."
- 1990—Shujian and his team transplanted cell nuclei of the mutant cell line (AHZC-88), which was resistant to the grass carp hemorrhagic virus, into unfertilized grass carp eggs using electric fusion, and raised three of the fish to the fry stage.
- 1999—Gong and his team produced transgenic Zebrafish by inserting green fluorescent protein (GFP), originally extracted from a jellyfish, that naturally produced bright green bioluminescence.
- 2003—Gong and his team developed transgenic zebrafish (*Danio rerio*) for ornamental and bioreactor system by strong expression of fluorescent proteins in the skeletal muscle.

Reference

Kaushik S, Kaushik S, Sharma D (2019) Functional genomics. Encyclopaedia Bioinform Comput Biol 2:118–113

Nucleic Acids

2

Nucleic acids consist of deoxyribonucleic acid (DNA) and ribonucleic acid (RNA) which was first discovered by Friedrich Miescher in 1869. DNA are the macromolecules considered the hereditary material that stores the genetic information of all living organisms.

2.1 Components of Nucleic Acid

Nucleic acids are the polymer molecules of nucleotides made up of three essential components: a heterocyclic nitrogenous base, a pentose sugar, and a phosphate group.

2.1.1 Nitrogenous Bases

Nitrogenous bases can be classified, viz. monocyclic bases called pyrimidines which comprises of a hexagonal aromatic ring and bicyclic bases called purines which comprises of one hexagonal and one pentagonal aromatic ring. Purines are adenine (A) and guanine (G), and pyrimidines are cytosine (C), thymine (T), and uracil (U) (Fig. 2.1). Both DNA and RNA contain adenine, guanine, and cytosine. However, thymine is found only in DNA which is replaced by uracil in RNA.

2.1.2 Pentose Sugar

In DNA, the pentose sugar is 2′ deoxyribose sugar, whereas in RNA, it is ribose which is the cyclic five carbon sugar (Fig. 2.2).

C. J. Betsy, C. Siva, *Fisheries Biotechnology and Bioinformatics*,
https://doi.org/10.1007/978-981-99-6991-3_2

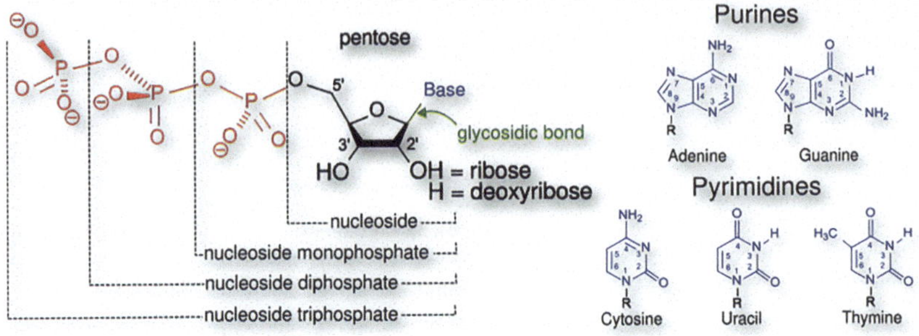

Fig. 2.1 Chemical structure of purines and pyrimidines

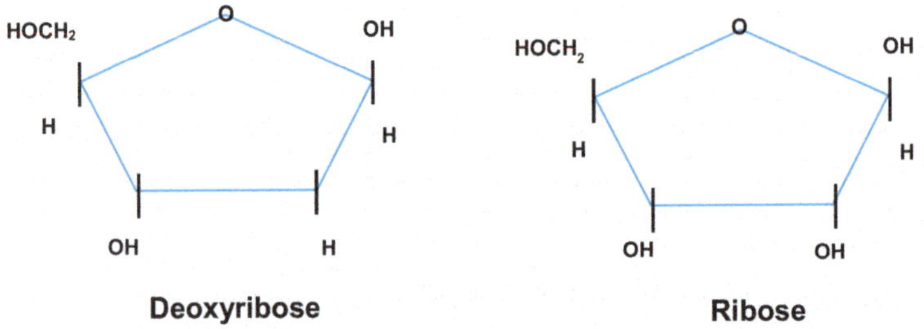

Fig. 2.2 Chemical structure of pentose sugar (ribose/deoxyribose)

2.1.3 Phosphoric Acid

The phosphate group is attached to 5′ carbon of pentose sugar molecule (ribose/deoxyribose) by phosphodiester linkage. This group is responsible for the strong negative charge on both the nucleotides and nucleic acids.

2.2 Nucleoside and Nucleotide

Nucleoside contains a nitrogenous base with a pentose sugar molecule while nucleotide contains pentose sugar molecule and phosphate group (Fig. 2.3).

2.3 Polynucleotide or Nucleic Acid Strand

The polynucleotide strand is made of several repeating units called nucleotides consisting of nitrogenous bases which are capable of being covalently linked together to form a long chain. The 3′-hydroxyl group on the ribose unit of the first nucleotide/deoxyribonucleotide reacts with the 5′-phosphate group (phosphodiester bond) on its neighbor to form a chain. Further, purine or pyrimidine is linked to the sugar by a glycosidic bond between a nitrogen and the 1′ carbon of the deoxyribose sugar (Fig. 2.4).

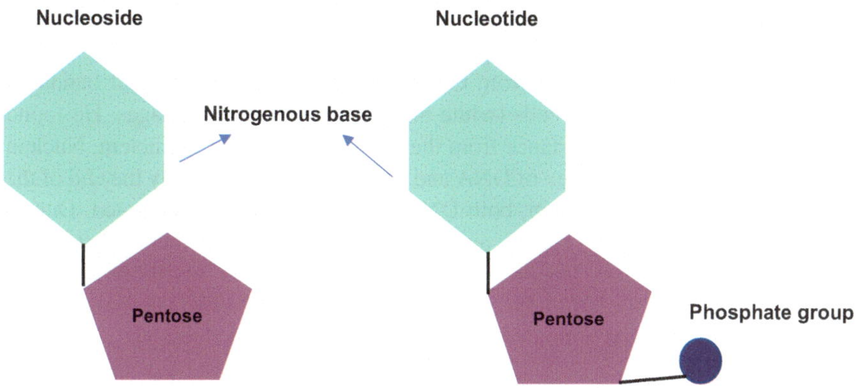

Fig. 2.3 Structure of nucleoside and nucleotide

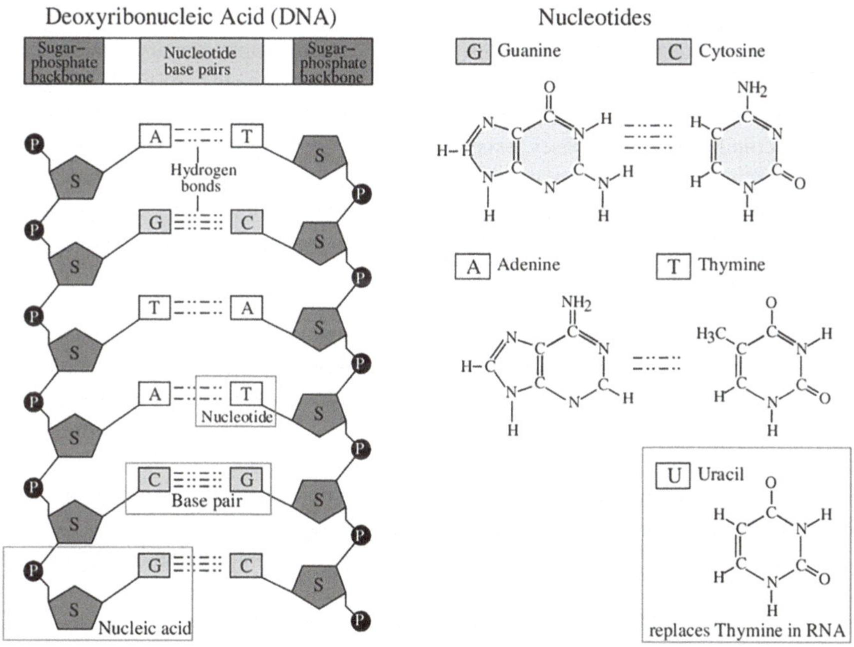

Fig. 2.4 Formation of polynucleotide (reproduced with permission from National Human Genome Research Institute)

2.4 DNA as Genetic Material

DNA contains the hereditary information. A gene is a DNA segment containing genetic code which are the biochemical instructions that specify the chemical structure of a particular protein. In a protein, the arrangement of amino acids is determined by the DNA structure of a gene which decides the protein type and function.

In 1869, Friedrich Miescher from Germany separated waste surgical bandages and isolated nuclei from pus cells (white blood cells) of those bandages. He found new phosphorus bearing substance from the nuclei and called it as nuclein. Nuclein is mostly chromatin, a complex of DNA and chromosomal protein. By the end of the nineteenth century, from protein, both DNA and RNA had been separated. During early 1930s, P. Levene, W. Jacobs, and others demonstrated the chemical structure of RNA and DNA. They reported that RNA is composed of ribose and four nitrogenous bases while DNA has deoxyribose and four bases. They also found that each base is coupled with a sugar-phosphate to form a nucleotide.

2.4.1 DNA Structure

In general, the structure of DNA is proposed to be a twisted ladder due to the winding of two chains of nucleotides. There are two complementary strands in a double helix structure. Sugars and phosphates constitute the sides of the ladder, and the rings are made up of bonded pairs of nitrogenous bases such as A, G, C, and T. RNA contains the same bases, except that U replaces T. Adenine and guanine are purines with two ringed structures. Thymine and cytosine are pyrimidines with single ringed structures. The presence of hydrogen bonds determines the specificity of base pairing. Adenine and thymine are aligned by two hydrogen bonds, while cytosine and guanine are formed by three hydrogen bonds. These are called base pairs.

The two chains unwind, and free nucleotides attach to the exposed bases of the now-separated chains, if the hydrogen bonds are broken. Each base is attached to a sugar and a phosphate molecule and is called as nucleotide. Double helix structure is formed because nucleotides are arranged in two long strands to form a spiral that resembles a ladder. DNA can replicate and hence each strand of DNA serves as a pattern for duplicating the base sequences. It is important during cell division as every new cell must have same copy of DNA present in old cell.

DNA strands have directionality due to the chemical composition of the pentose residues. In a DNA strand, an exposed hydroxyl group is present on the deoxyribose in one end of a DNA polymer and is known as the 3′ end while an exposed phosphate group is present on the other end which is the 5′ end. The two strands of a double-helix run in opposite directions. Nucleic acid synthesis, including DNA replication and transcription occurs in the $5' \rightarrow 3'$ direction, because new nucleotides are added via a dehydration reaction that uses the exposed 3′ hydroxyl as a nucleophile.

2.4.1.1 The Double Helix

Linus Pauling elucidated the helix for DNA structure. Watson and Crick proposed the double helix model of DNA. The 3D structure of the DNA was analyzed by

Maurice Wilkins, and Rosalind Franklin used X-ray diffraction which confirmed that DNA was a helix (Fig. 2.5).

Erwin Chargaff studied the base composition of DNA and gave Chargaff's rules which provided a valuable confirmation of Watson and Crick's model,

- The content of purines always equalled the content of pyrimidines.
- The amounts of adenine and thymine; guanine and cytosine were always equal.

The important features of double helix are given below:

- The best model that satisfied all the X-ray data was a double helix with the sugar phosphate chain on the outside and the bases on the inside.
- Alternating phosphate and sugar molecules form each side of this ladder.
- Bases from one DNA strand join with bases from another strand to form the rungs of the ladder, holding the double helix together.
- Adenine always pairs with thymine, and guanine always pairs with cytosine.
- Complementary base pairing enables to predict the base sequences of one DNA strand if the order on the corresponding or complementary DNA strand is known.
- The two chains run in an anti-parallel fashion with one chain having a $5' \rightarrow 3'$ orientation and the other having a $3' \rightarrow 5'$ orientation.
- DNA sequences are read from $5' \rightarrow 3'$ with respect to the polarity of the strand.
- Helical structure is right handed.
- The width of the helix was found to be 2 nm, and it can accommodate only two strands.
- The purine and pyrimidine bases were stacked 0.34 nm apart in a ladder.

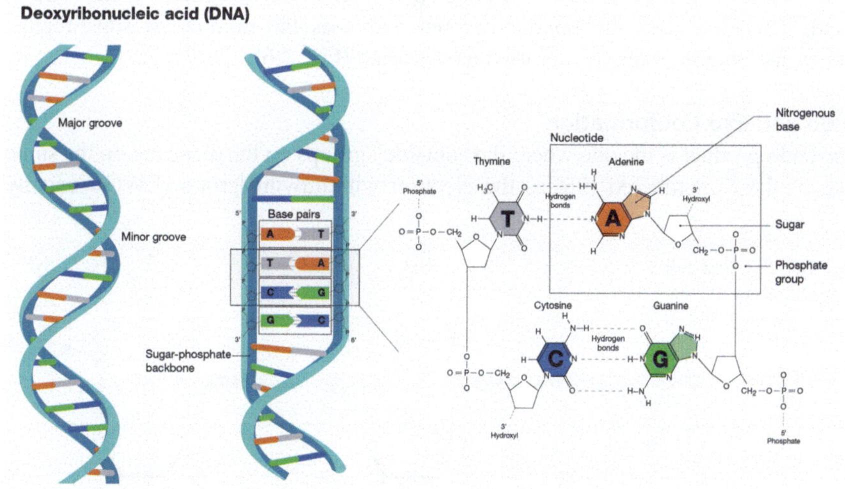

Fig. 2.5 Structure of DNA double helix (reproduced with permission from Darryl Leja, National Human Genome Research Institute)

- The helix made one full turn every 3.4 nm and, therefore, there should be ten layers of bases stacked in one turn.
- Each step would contribute a pair of bases, with each base attached to one of the sugar-phosphate backbone.
- In a given DNA, adenine is equal to thymine and guanine to cytosine.
- A and T pair with two hydrogen ponds, whereas G and C pair with three hydrogen bonds.
- The fifth (5-prime, of 5′) carbon of the pentose ring is connected to the third (3-prime, of 3′) carbon of the next pentose ring via a phosphate group, and the nitrogenous bases stick out from this sugar-phosphate back bone.

2.4.1.2 Minor and Major Grooves

The backbones of the strand are closer on the one side of the helix than on the other. The stacking of the bases occurs when the phosphodiester backbone faces outward and nitrogenous bases face inward leading to the formation of different grooves on opposite sides of the base pairs called major and minor grooves. When there is more distance between the backbones, major groove occurs and when the backbones are closer, the minor groove occurs. The grooves twist around the molecule on opposite sides. The major groove is approximately 50% wider than the minor.

2.4.1.3 Conformations

It is the spatial arrangement which the atoms in a molecule may adopt and freely convert between, especially by rotation about individual single bonds. It includes the following types.

Syn Conformation

In syn addition, two substituents are added to the same side of a double bond or triple bond. This decreases the bond order but increases the number of substituents. Generally, alkene or alkyne are used as substrate (Fig. 2.6).

Endo and Exo Conformation

The endo product is the one where the "outside" groups on the diene are on the same side of the 6-membered ring as the electron withdrawing group (EWG—usually

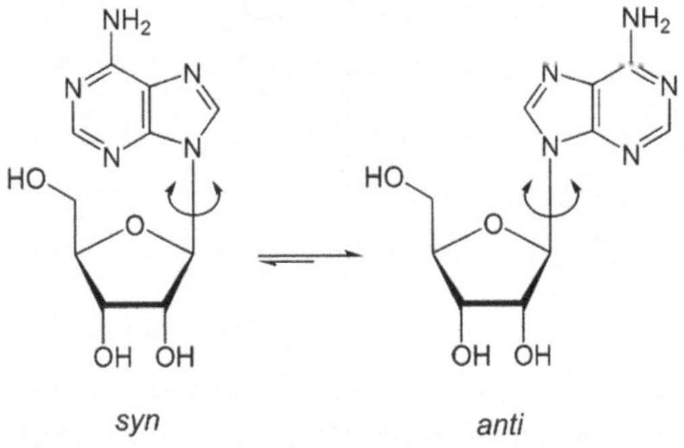

Fig. 2.6 Syn and anti-conformations

contains a C=O or CN group). The exo product is the one where the "outside" groups are on the opposite side of the ring as the 6-membered ring (Figs. 2.7, 2.8, and 2.9).

2.4.2 Different Forms of DNA

The double helix form of DNA is called B-DNA. However, DNA doesn't always have enough space to form a perfect, linear B-DNA conformation. In all form of cells, DNA must be folded thousand times in order to fit inside the nucleus which

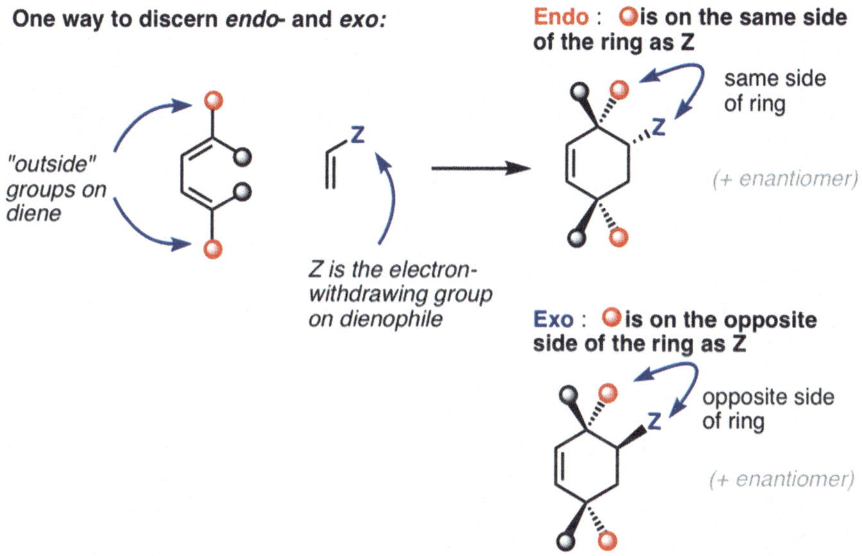

Fig. 2.7 Exo vs Endo products in the Diels Alder (reproduced with permission from Dr. James Ashenhurst, Master Organic Chemistry)

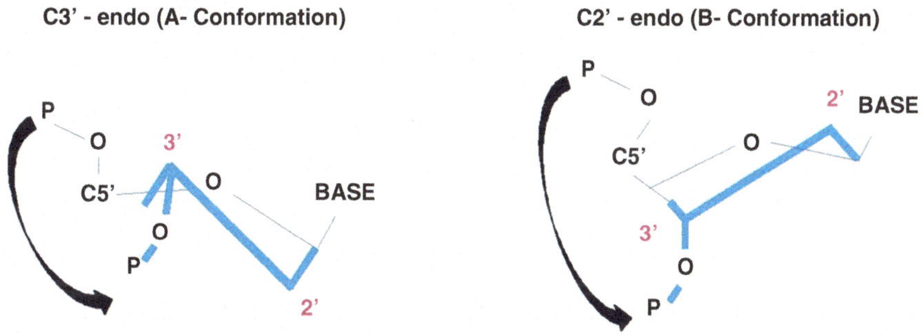

Fig. 2.8 Sugar puckers in DNA/RNA structures with A- or B-form conformation

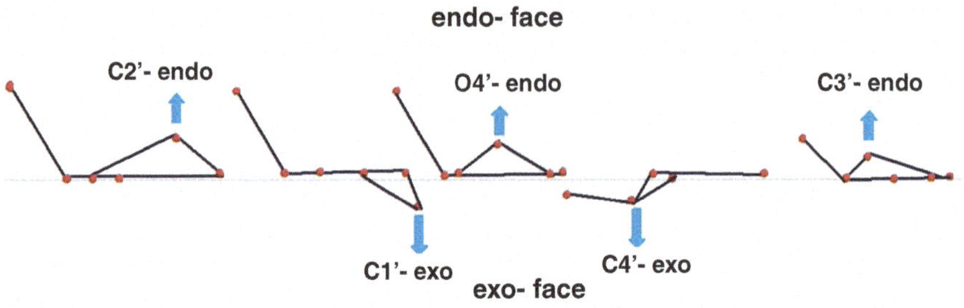

Fig. 2.9 Sugar pucker

Table 2.1 Comparisons of B-form, A-form, and Z-form DNA

	B-Form	A-Form	Z-Form
Helix sense	Right handed	Right handed	Left handed
Base pairs per turn	10	11	12
Vertical rise per bp	3.4 Å	2.56 Å	19 Å
Rotation per bp	+36°	+33°	−30°
Helical diameter	19 Å	19 Å	19 Å
Axial rise	–	0.26 nm	0.45 nm
Helix pitch	34°	28°	45°
Base pair tilt	−6°	20°	7°
Twist angle	36°	33°	−30°
Helix diameter	2.0 nm	2.3 nm	1.8 nm

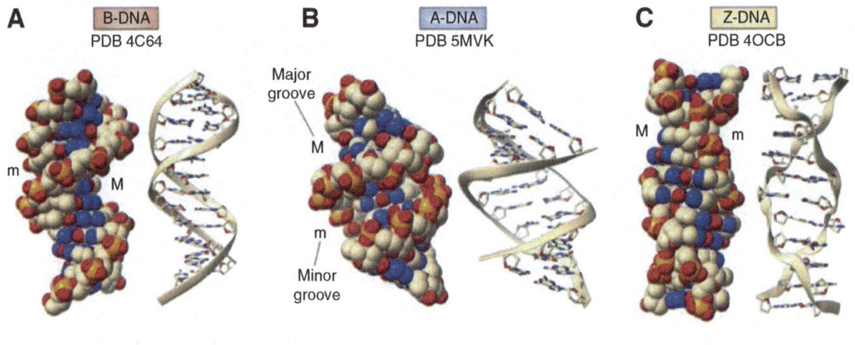

Fig. 2.10 Crystal structures of B, A, and Z form double helices (reproduced from Neidle 2021)

leads to different forms of DNA such as A-DNA, B-DNA, and Z-DNA (Table 2.1; Fig. 2.10).

2.4.2.1 B-DNA

- It is the original double helix form of DNA as proposed by Watson–Crick.
- The two strands of DNA wound around the same axis and are antiparallel and plectonemically coiled.

- The nucleotides in one strand in a 5′–3′ orientation align with nucleotides on another strand in the 3′–5′ orientation.

2.4.2.2 A-DNA

- The confirmation of the deoxyribose sugar ring and placement of base-pairs within the duplex determines the A-form and B-form DNA.
- It is in the C2′ endo conformation for B-form, whereas for A-DNA, it is in the C3′ endo conformation.
- The base-pairs are centered over the helical axis in B-DNA, whereas base-pairs are displaced from center of the helical axis and are placed closer to the major groove in A-DNA resulting in a ribbon-like helix with an open cylindrical core in A-DNA.

2.4.2.3 Z-DNA

- It was discovered by Rich, Nordheim, and Wang in 1984.
- In Z-DNA, the two strands coil in a zig-zag pattern in the phosphodiester backbone.
- It forms when the DNA is in an alternating purine-pyrimidine sequence such as GCGCGC.
- The difference is in the G nucleotide which has the sugar in the C3′ endo conformation, and the guanine base is in syn-conformation which places the guanine over the sugar ring.
- The duplex in Z-DNA has to accommodate the distortion of this G nucleotide in the syn-conformation. The cytosine in the adjacent nucleotide of Z-DNA is in the "normal" C2′ endo, anti-conformation.

2.4.2.4 Conditions Favoring A-Form, B-Form, and Z-Form of DNA

The following conditions decide the A-, B-, or Z-DNA conformation:

- Ionic or hydration environment: This can facilitate conversion between different helical forms. A-DNA is formed by low hydration, whereas Z-DNA is formed by high salt.
- DNA sequence: A-DNA is formed by stretches of purines or pyrimidines, while Z-DNA is formed by alternating purine-pyrimidine.
- Presence of proteins: Proteins can bind to DNA in one helical conformation and force the DNA to adopt a different conformation. For instance, proteins can bind to B-DNA and can force it to either A- or Z.

2.5 Ribonucleic Acid (RNA)

In the process of conversion of DNA into protein, RNA is the first intermediate formed. RNA is single-stranded with four nucleotides A, C, G, and U joined by alternating phosphate and ribose sugar residues in the backbone. Some RNAs are directly involved in cellular metabolism. RNA is produced by copying the base sequence of certain section of DNA, called gene, into a piece of single-stranded nucleic acid by a process called transcription which is catalyzed by an enzyme, RNA polymerase.

2.5.1 Chemical Structure of RNA

RNA is sensitive to oxidizing agents. In double-stranded DNA, the bases in one strand pair with bases in another strand. In contrast to this, single-stranded RNA has complex three-dimensional structures since the bases pair with other bases of the same strand. Sometimes, intermolecular RNA/RNA duplexes also form. Since RNA are flexible molecules, they form variety of structures through internal base pairing and additional non-base pair interactions. They can form even hairpin loops such as in transfer RNA (tRNA) (Fig. 2.11).

2.5.2 Types of RNA

2.5.2.1 Messenger RNA (mRNA)

Functional part of DNA which plays an important role in protein synthesis.

2.5.2.2 Transfer RNA (tRNA)

Act as adapters which carry specific amino acids from the cytoplasm on to the ribosomes during synthesis of proteins.

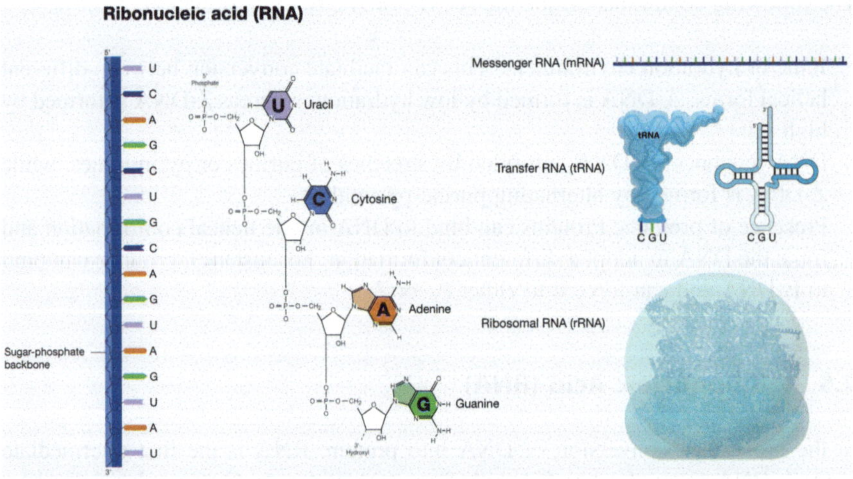

Fig. 2.11 Structure of RNA and its different types (reproduced with permission from Darryl Leja, National Human Genome Research Institute)

2.5.2.3 Ribosomal RNA (rRNA)

RNA component of the ribosome, the site of protein synthesis in all living cells.

2.6 Difference Between DNA and RNA

The difference between DNA and RNA in terms of structure and function is provided in Table 2.2 and depicted in Fig. 2.12.

Table 2.2 Comparison between DNA and RNA

Comparison	DNA	RNA
Full name	Deoxyribonucleic acid	Ribonucleic acid
Function	Replicates and stores genetic information	Converts the genetic information and moves it to ribosomal protein factories
Structure	Double stranded, arranged in a double helix, made up of nucleotides	Single stranded, made up of nucleotides. RNA strands are shorter than DNA strands
Length	Longer than RNA	Shorter than long DNA polymers
Sugar	Deoxyribose	Ribose
Bases	Adenine ("A"), thymine ("T"), guanine ("G"), and cytosine ("C")	Adenine ("A"), uracil ("U"), guanine ("G"), and cytosine ("C")
Base pairs	A-T; C-G	A-U; C-G
Location	Nucleus	Nucleolus
Reactivity	More stable molecule than RNA	More reactive than DNA and is not stable in alkaline conditions. The larger helical grooves of RNA mean that it is prone to enzyme attack
Ultraviolet (UV) sensitivity	Vulnerable to damage by UV light	More resistant

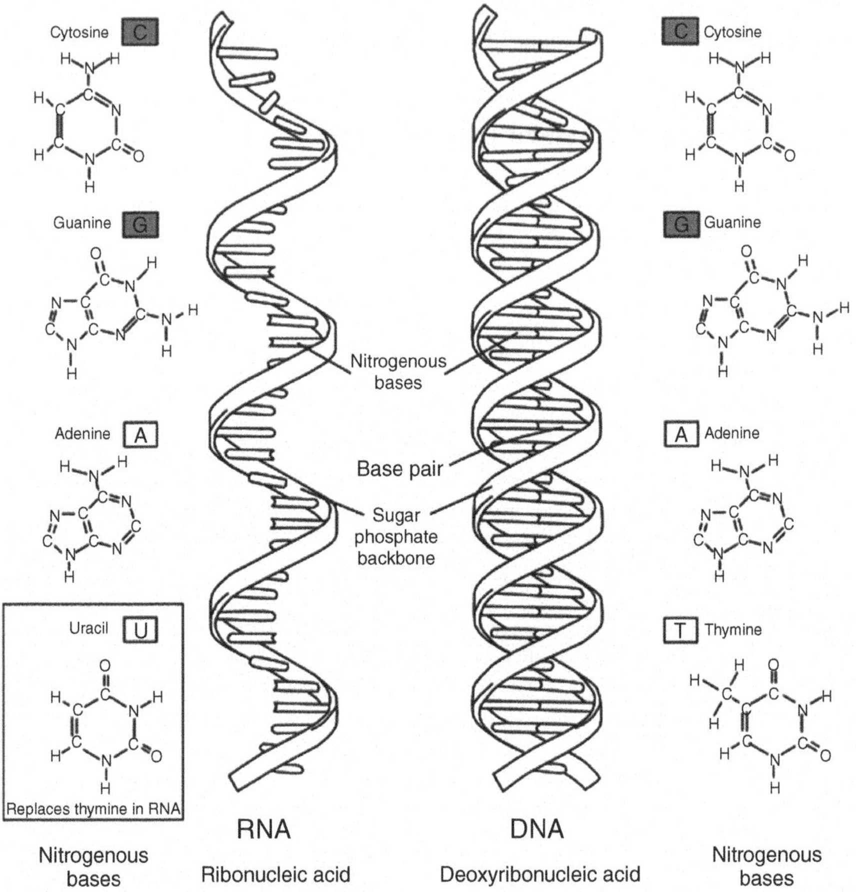

Fig. 2.12 Structural difference between DNA and RNA (reproduced from Noreen et al. 2020 with permission from Elsevier)

References

Neidle S (2021) Beyond the double helix: DNA structural diversity and the PDB. J Biol Chem 296: 100553

Noreen A, Sultana S, Sultana T, Tabasum S, Zia KM, Muzammil Z, Jabeen M, Zaeem A, Sultana S (2020) Natural polymers as constituents of bionanocomposites. In: Bionanocomposites, pp 55–85. https://doi.org/10.1016/B978-0-12-816751-9.00003-9

Genes and Genetic Codes

3

3.1 Genes

Genes are the basis of inheritance and are transferred from parents to offspring, and traits are inherited. Genes are arranged within chromosomes which has the single, long DNA molecule, a portion of which corresponds to a single gene. A gene could be as short as a few hundred base pairs or as long as many thousands. The term gene was introduced in 1909 by Danish geneticist, Wilhelm Johannsen. It was derived from the ancient Greek word, γόνος, *gonos*, which means offspring and procreation.

In a total genome, the genes can be classified as functional and non-functional genes. In humans, functional genes contribute to only 3% of the total genome and the remaining portion of 97% consists of "junk DNA." Some of this junk DNA consist of pseudogenes, which becomes non-functional genes. The non-coding DNA consists of dispersed or clustered repeated sequences of varying length, from one base pair (bp) to thousands of bases (kilobases, kb) in length.

- Dispersed repeated sequences—They occur as copies which spread across the genome and are categorized as long or short interspersed nuclear elements, long terminal repeats, and DNA transposons.
- Clustered repeated sequences—The repeated sequences occur in tandem copies and depending on the length of the repeat unit, they are called as satellites, minisatellites, and microsatellites. Repeated elements can constitute up to 40% of the genome.

3.1.1 Activities of Genes

- A gene can be replicated and thus genetic information can be passed from generation to generation unchanged.

C. J. Betsy, C. Siva, *Fisheries Biotechnology and Bioinformatics*,
https://doi.org/10.1007/978-981-99-6991-3_3

- The sequences of bases in the RNA depends directly on the sequences of bases in the gene. Most of these RNAs serve as templates for making protein molecules. Thus, most genes are essentially blueprints for making proteins. The production of protein from a DNA blueprint is called gene expression.
- A gene can accept occasional changes or mutations.

3.1.2 Structure of Gene

A gene is a small portion in a DNA which is held as a code. This code specifies the amino acid sequence of a protein. The sequence information for a single gene will not be continuous along the DNA, but will be interspersed with pieces of non-coding sequence. The coding parts of a gene sequence are exons, and the non-coding parts are introns. Genes are DNA strands made up of the nucleotide chain. The chemical structure of a gene comprises of nucleotides. The nucleotide of one strand binds with the nucleotide of the opposite strand by hydrogen bonding, whereas adjacent nucleotides bind with each other by phosphodiester bonds. Gene structure consists of core elements and regulatory elements. The core elements like exons are involved in protein formation, and the regulatory elements such as promoters, enhancers, and silencers maintain gene expression. Another type of element, the maintenance element contains information for DNA repair, DNA modification, and DNA replication. The functional structure of a gene comprises of introns, exons, promoters, enhancer, and untranslated region. In a given gene, after splicing, the coding parts are joined and are called exons, whereas the non-coding sequences that are removed from the final transcript are called introns. The elements located at the end of a gene are called regulatory elements (Fig. 3.1).

The non-coding sequences that facilitate binding sites for enzymes and transcriptional factors are called promoters. The promoter is located near 5′ end, and it consists of TATA box and CCAAT sequences for enzyme binding. The promoter is made up of

- Core promoter—helps in RNA polymerase bindings to initiate transcription
- Proximal promoter sequence—provides bindings for transcriptional factors

The induction and repression of transcription are by enhancer and silencer, respectively, and are located away from exon to regulate gene expression. The 3′ untranslated regions are non-coding regions of gene helping in aborting the process of transcription and forming the final transcript. Once the RNA polymerase reaches the untranslated region, it stops synthesizing RNA and detached from the strand.

3.1.3 Types of Genes Based on Their Function and Position

Housekeeping genes: They are required to perform normal cell functions and generally code for transcription, translation, and replication.

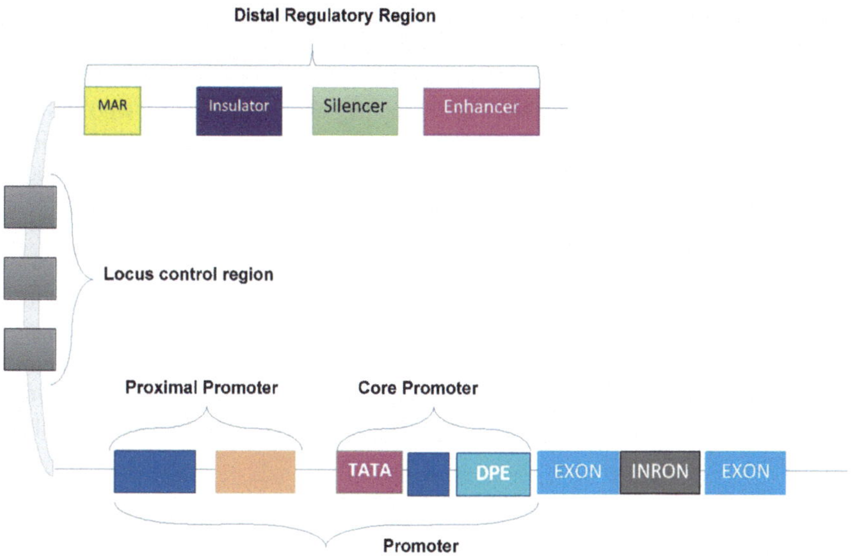

Fig. 3.1 The molecular structure of a gene

Inducible genes: They remain inactive and is expressed under the influence of extrinsic factors.

Developmental genes: They are involved in the early development of organisms.

Tissue-specific genes: They are active only in specific tissues and remain inactive in other tissues.

Homologous genes: They are inherited from a common ancestor, share a common function, and have sequence similarities and are categorized into homologous genes.

Non-homologous genes: They aren't inherited from a common ancestor; instead, it is originated due to some evolutionary forces.

Autosomal genes: They are located on autosomal chromosomes.

Sex-linked genes: They are located on sex chromosomes.

3.2 Genetic Code

The set of instructions that translates DNA into 20 amino acids are called genetic code. Codons, the three letter nucleotides are the basis of genetic code. Every codon codes only for one specific amino acid. The genetic code determines the amino acid sequence in a protein.

3.2.1 Discovery of the Genetic Code

Codon was discovered by Francis Crick and his colleagues during 1961. Later, the genetic code was deciphered by Marshall Nirenberg and his colleagues who showed that during protein synthesis, A, T, G, C form codons of different base combinations which code for all 20 amino acids. Following this, Nirenberg along with Johann Matthaei, a German scientist conducted various experiments to know about protein synthesis using synthetic RNA.

They used a "cell-free system" to add RNA strands with any of the four bases (A or G or U or C) and radioactively tagged amino acids. Using radioactive measurements, they found that when RNA containing the base U was added, molecules with one single amino acid, phenylalanine, were synthesized. Hence the triplet made up of UUU resulted in phenylalanine. Following this method, by 1963, scientists deciphered 35 codons and by 1966, they had deciphered more than 60 codons.

3.2.2 Codons and Amino Acids

In the genetic code, each codon consists of three bases which are arranged in a specific order. Each codon combination corresponds to one particular amino acid. It is known that from the four nucleotides, 64 possible permutations and combinations of three letter nucleotide sequences are possible. In total, there are 64 codons, 61 codons code for different amino acids, and three codons are considered stop codons.

3.2.2.1 START and STOP Codons

START codon begins the translation. The common start codon is AUG. In eukaryotes, AUG codes for methionine while in prokaryotes, it codes for formyl methionine. There are three STOP codons in total. They are also called as nonsense or termination codons. Codons such as UAG, UGA, and UAA are considered STOP codons and are named amber, opal, and ochre, respectively. During protein synthesis, they signal the end of the polypeptide chain. They also trigger the ribosome to release the new polypeptide chain since no tRNA anticodons complement these stop codons (Fig. 3.2).

3.2.3 Properties of the Genetic Code

1. The code is degenerate or redundant. Most of the amino acids are coded for more than one codon. Leucine, serine, and arginine have six different codons. Proline, threonine, and alanine have four. Isoleucine has three. Methionine and tryptophan have only one codon.
2. The code is unambiguous. One codon codes for one amino acid only.

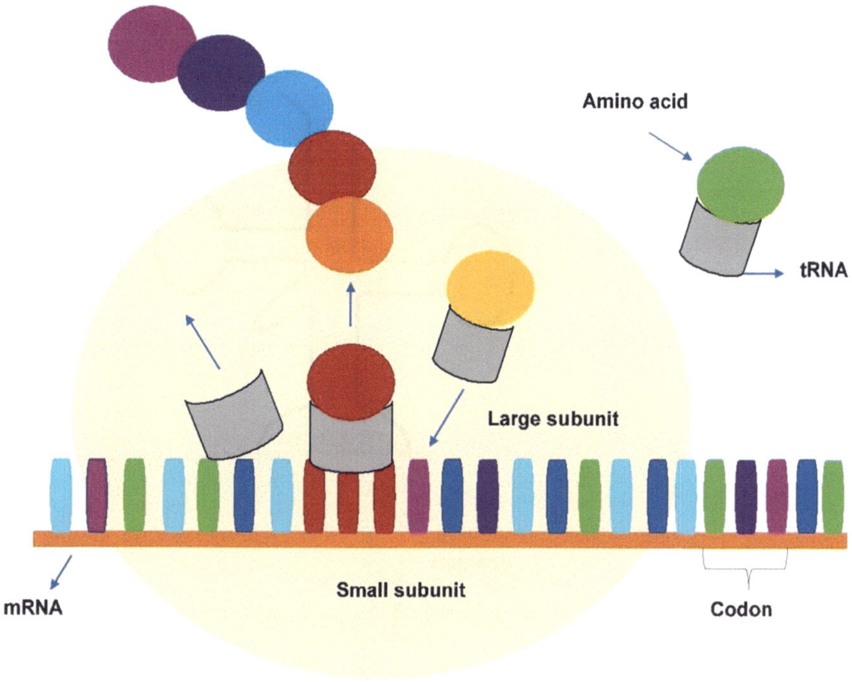

Fig. 3.2 Process of amino acid formation in ribosome from codon

3. The code is highly universal. It is the same for various different kind of organisms. However, a few exceptions to this are known as in yeast mitochondria, UGA codes for tryptophan instead of stop. In Paramecium, UAA and UAG code for glutamine instead of stop codon.

3.2.4 The Wobble Hypothesis

The property of the genetic code is that the code is unambiguous. However, sometimes, more than one codon code for one amino acid and is known as degeneracy of genetic code. In order to explain this degeneracy of codons, Francis Crick proposed "Wobble hypothesis" in 1966. The term wobble means to sway or move unsteadily. In this hypothesis, he suggested that among the three bases, the first two bases of the codon have an exact pairing with the bases of the tRNA anticodon. There may be wobbling when pairing occurs between the third base of the codon and anticodon (Fig. 3.3). This leads to the recognition of more than one codon by a single tRNA. Hence even if there are 61 codons for amino acids, there will be around 40 tRNA only due to wobbling.

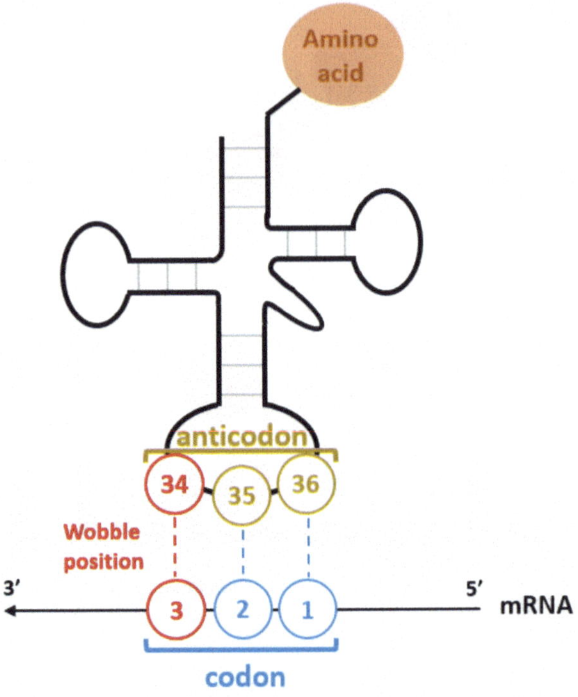

Fig. 3.3 Codon-anticodon interaction (reproduced from Fages-Lartaud and Hohmann-Marriott 2022)

According to wobble hypothesis, the base present at 5′ end of the anticodon is not spatially confined as the other two bases and hence it forms hydrogen bonds with any of the bases located at the 3′ end of a codon. This leads to the following conclusions:

- In the codon, the first two bases have normal (canonical) hydrogen bond pairs with the second and third bases of the anticodon.
- Non-canonical pairing may occur in the remaining position. Therefore, at the third position of the codon, the wobble hypothesis proposes a flexible set of base pairing rules.
- The wobble allows the anticodon of one tRNA to pair with more than one triplet in mRNA.
- According to the wobble base pair rules, first base U can recognize A or G, first base G can recognize U or C, and first base I can recognize U, C, or A.

3.2.4.1 Wobble Base Pairs

The pairing that doesn't follow Watson—Crick base pair rules are known as wobble base pair (Fig. 3.4). The four main wobble base pairs are guanine-uracil (G-U), hypoxanthine-uracil (I-U), hypoxanthine-adenine (I-A), and hypoxanthine-cytosine (I-C). Hypoxanthine is the nucleobase of inosine; hence "I" is used for hypoxanthine.

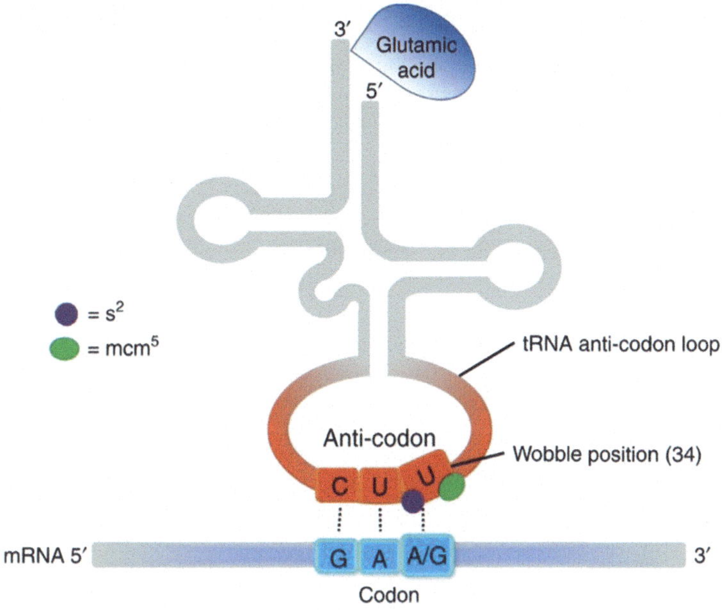

Fig. 3.4 Wobble position (reproduced from Goffena et al. 2018)

3.2.4.2 Significance of the Wobble Hypothesis

- Wobble provides broad specificity with limited amount of tRNAs.
- In RNA secondary structure, wobble base pairs are essential and are necessary for proper translation of the genetic code.
- Wobbling allows faster dissociation of tRNA from mRNA and also protein synthesis.
- The existence of wobble minimizes the damage that can be caused by a misreading of the code; for example, if the Leu codon CUU was misread CUC or CUA or CUG during transcription of mRNA, the codon would still be translated as Leu during protein synthesis.

References

Fages-Lartaud M, Hohmann-Marriott MF (2022) Overview of tRNA modifications in chloroplasts. Microorganisms 10(2):226. https://doi.org/10.3390/microorganisms10020226

Goffena J, Lefcort F, Zhang Y et al (2018) Elongator and codon bias regulate protein levels in mammalian peripheral neurons. Nat Commun 9:889. https://doi.org/10.1038/s41467-018-03221-z

Mutations and Their Implications 4

In general, genetic diversity initially occurs as a result of mutations. The genetic material of any organism is either DNA or RNA, and any change in the DNA leads to genetic variations. The uniqueness of individuals within a species is due to DNA and sexual reproduction. In principle, genetic information flows from DNA to RNA to protein. The coding region of the DNA forms the protein. The protein's primary structure is determined by sequences of nucleotides or bases that code specific amino acid sequences.

	DNA	Transcription →	mRNA	Translation →	Protein
NORMAL	3'-CTC-5'		5'-GAG-3'		Glu
MUTATION	3'-CAC-5'		5'-GUG-3'		Val

Hence mutations are mistakes in a cell's DNA that leads to abnormal protein production. These changes may be due to internal or external mechanisms or specific agents and are called mutations.

4.1 Effects of Mutations

Mutations are rare and can be harmful, neutral, or helpful.

- Harmful mutations—Affect the survival of the organism. The individual dies before they can reproduce; thus, the mutant gene is eliminated.

C. J. Betsy, C. Siva, *Fisheries Biotechnology and Bioinformatics*, https://doi.org/10.1007/978-981-99-6991-3_4

- Neutral mutations—They neither help nor hinder the individual's survival. The individual reproduces and passes on the neutral mutated gene to the next generation.
- Helpful mutation—It helps the individual to survive in the environment.

4.2 Types of Mutations

- Gene mutations.
- Chromosomal mutations.

4.2.1 Gene Mutations

The gene mutations are classified as follows:

(a) Spontaneous mutations.
(b) Insertion and deletion mutations or frameshift mutations.
(c) Transposons.

4.2.1.1 Spontaneous Mutations

The spontaneous mutation is due to definite biochemical changes. It occurs as a result of any of the following internal factors:

- Replication error of DNA
- Mistake in the recombination
- Mis-pairing of DNA damage
- Depurination
- Deamination of the bases
- Movement of transposons

These are further classified as follows:

1. Base pair substitution
2. Silent mutations
3. Neutral mutations
4. Missense mutations
5. Nonsense mutations (Amber mutations)

Base Pair Substitution

As the name suggests, in this type of gene mutation, base pairs are substituted like purine to purine, pyrimidine to pyrimidine, and pyrimidine to purine or vice versa. For example, if G is present in a strand, then C should be present in a complementary strand. Due to mutation, if G is replaced by A, then instead of GC combination, AT will be formed. Base pair substitution is further classified into

- Transition mutation—purine to purine or pyrimidine to pyrimidine replacement occurs in the same strand of DNA. E.g., GC is replaced by AT and AT is replaced by GC.
- Transversion mutation—pyrimidine to purine or purine to pyrimidine replacement occurs on the same strand of DNA. E.g., GC to CG or TA and AT to TA or GC.

Silent Mutations

Mutations that do not always produce visible phenotypic changes are known as silent mutations. E.g., if in a codon CUU due to mutation, it becomes CUA or CUG or CUC, it will code for amino acid leucine.

Neutral Mutation

Neutral mutations are also base pair substitutions in the allele's codon. Though the codon produces a different amino acid, the change of a few amino acids in the primary structure does not change the protein's function. For example, if in the original allele's codon is CUU, the codon CUU will code leucine. Due to mutation, if AUU replaces CUU, the amino acid isoleucine will be coded. The two amino acids, leucine and isoleucine, are chemically similar. Hence the change in amino acid would not alter the protein's function; thus, there will be no phenotypic change.

Missense Mutation

Missense mutation occurs where there is a substitution in only one base pair resulting in the formation of a new amino acid. Sometimes it causes some diseases. Hypertrophic cardiomyopathy in human beings is caused due to missense mutations in exon 13 of the MHC (Myosin heavy chain) β chain resulting in the change of adenine for guanine and the formation of glutamine instead of arginine. This missense mutation causes enlargement of heart (left ventricle).

Nonsense Mutation (Amber Mutations)

It is a mutation in which the substitution of base pairs results in the codon UGA, UAA, or UAG. These codons are nonsense codons or stop codons. No other amino acid is formed in such a mutation except the production of the original protein.

4.2.1.2 Frame Shift Mutations/Insertion and Deletion Mutations

In these mutations, there is insertion or deletion of one or two base pairs (not multiple of three) in DNA. This results in altered reading frame of mRNA. For example, if the DNA coding strand CAT CAT CAT CAT CAT has a single base pair deletion at base pair 6, the mRNA will read CAU CAC AUC AUC AUC, and so forth. The frame shift mutation usually has radical effect on the protein product.

4.2.1.3 Transposon Insertion

These are mobile elements in the genome and can jump and insert into the DNA. It stated that 1–10 kb DNA can move within the genome. It is also known that 50–80% of the mutations are caused by gene disruption. These are also responsible for genetic variation.

4.2.2 Chromosomal Mutations (Aberrations)

The difference between chromosomal and gene mutations is that the rearrangement involves long segments of DNA rather than single bases. It generally occurs at the time of DNA replication. Further recombination involves non-homologous sister chromatids (single DNA molecules from non-homologous chromatids) instead of sister chromatids.

The chromosomal theory of inheritance suggests that genes (DNA) are physically located on chromosomes and that Mendelian inheritance can be explained in terms of chromosome behavior during cell division. The chances of mutations are more and can be illustrated by the following example.

If the chromosome number in diploid organism is 10 pairs, then 10 are coming from male (sperm) and 10 are coming from female-ovum. Then the possible combinations would be $(2)^{10} = 1024$. These random combinations are possible according to Mendel's principle of the independence assortment, which means that large numbers of genetic variations are possible. The chromosomal mutation is a visible change in chromosome structure. Chromosomes themselves mutate and evolve.

Most of the chromosomal rearrangements arise as a result of a mistake during meiosis. Chromosome number for each species is fixed and in case if the number of chromosomes changes, it would be a new species. Sexual reproduction plays a prime role in the creation of genetic variations. The following are the types of chromosomal aberrations:

(a) Translocation
(b) Inversion
(c) Deletion
(d) Duplication

4.2.2.1 Translocation

In translocation, a chromosomal segment is moved from one position to another, either within the same chromosome or to another chromosome. It is considered balanced if there is no gain or loss in the genetic material (Fig. 4.1) and is known to

Fig. 4.1 Balanced Translocation

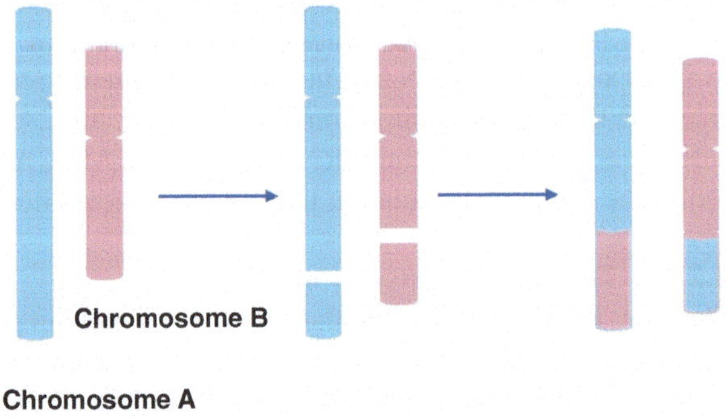

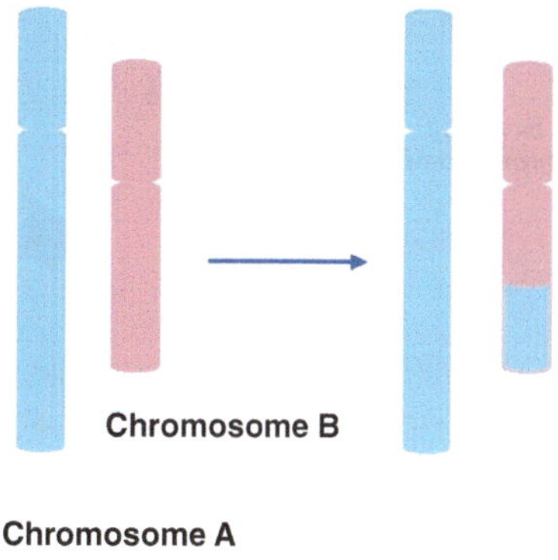

Fig. 4.2 Unbalanced Translocation

be unbalanced if there is a gain or loss in the genetic material (Fig. 4.2). In the Drosophila genus, the fusion of chromosomes and translocation between non-homologous chromosomes resulted in the formation of new species such as *subobscura, pseudoobscura, melanogaster, ananassae,* and *willistoni*. The fusion of chromosomes occurs when two non-homologous chromosomes fuse into one.

4.2.2.2 Inversion

In an inversion, a fragment of one chromosome breaks off and re-attaches to its original position in a reversed orientation. The original chromosome may contain the centromere as pericentric inversion (Fig. 4.3a), or it may not, as in paracentric inversion (Fig. 4.3b). The presence of loops in the cytological preparations of cell at the pachytene stage of meiosis can recognize chromosome heterozygote inversions.

4.2.2.3 Deletion

Chromosome deletions occur when the DNA strand breaks. The fragments that do not contain centromere (acentric fragments) will be lost during subsequent cell division (Fig. 4.4). Deletion leads to Cri-du-Chat syndrome in humans, in which mental retardation, growth restriction, and cat-like cry occur.

4.2.2.4 Duplication

Chromosomal duplication provides an additional copy of a block of DNA (pieces of the chromosome) having a complete gene sequence (Fig. 4.5). When duplication contains a complete gene sequence, natural selection can operate independently on both the new and the old sequence to produce divergent variants.

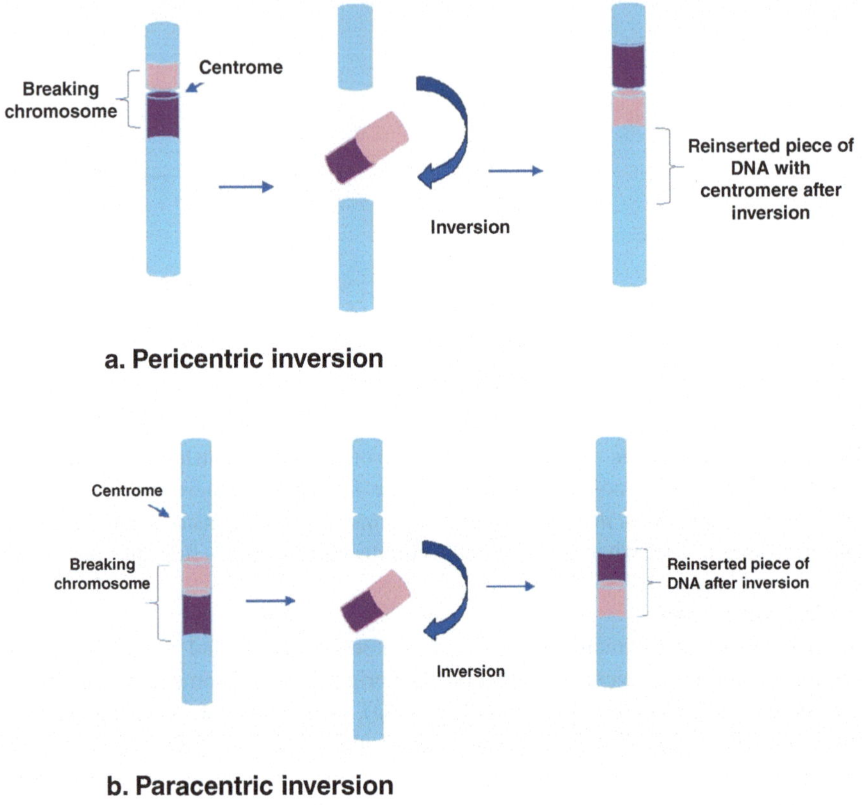

Fig. 4.3 Inversion (para and pericentric) of chromosome. (**a**) Pericentric inversion, (**b**) Paracentric inversion.

Fig. 4.4 Deletion of chromosome

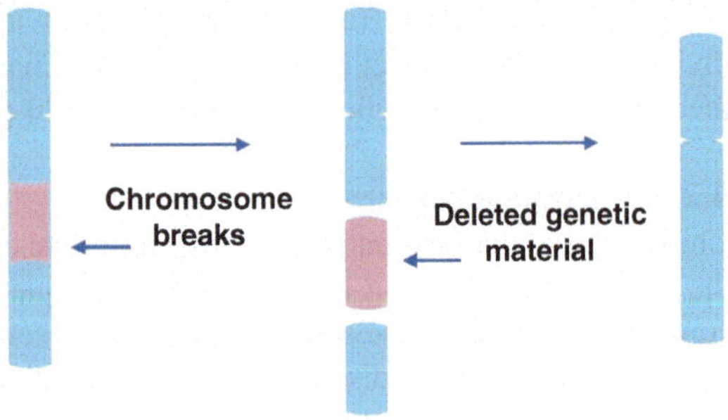

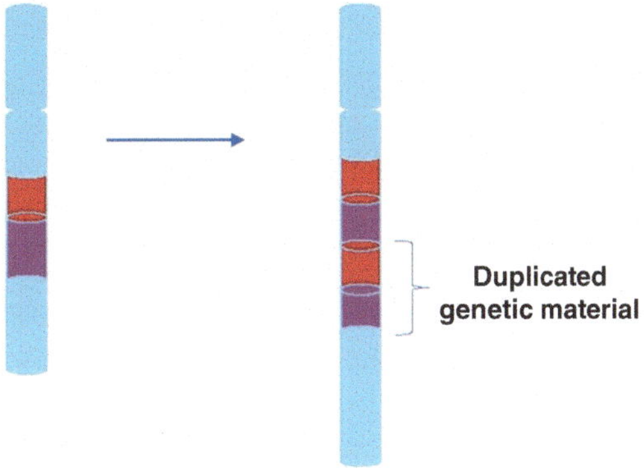

Fig. 4.5 Duplication of chromosome

4.2.2.5 Polyploidy

An increase in the number of chromosomes is known as polyploidy. It is a condition where individuals have more than two copies of each chromosome. For example, triploid has three sets of chromosomes, and tetraploid has four. Polyploidy occurs naturally in some plants, like hexaploid wheat. Tetraploidy is reported in salmonids. Polyploidy can be artificially induced in normally diploid species for aquaculture processes.

5 Gene Expression in Prokaryotes

Prokaryotes are single-celled organisms which doesn't have nucleus. Their DNA are free floating in the cell cytoplasm. In prokaryotes, proteins are synthesized by transcription and translation. The transcription stops when the resulting protein is not required which depends mainly on their immediate environment like presence and absence of nutrients. Hence, gene expression in prokaryotes occurs primarily at the level of transcription. Operons regulate the gene expression in prokaryotes like bacteria. François Jacob and Jacques Monod first identified operon in 1961. A cluster of functionally related genes controlled through shared operator is called operons. They consist of multiple genes which are grouped together with a promoter and an operator. The presence of operons is experimentally determined using northern blot, RT-PCR, and gene expression arrays (Ramos et al. 2013).

5.1 Formation of Operon

The formation and loss of operons is a dynamic process. Horizontal gene transfer (HGT) is one source of operon formation. According to the selfish operon model, the functional gene clusters originate primarily from HGT (Rocha 2008) where the genes present in an operon are homologous with genes present in other sequenced organisms. Genes with unidentifiable homologues outside a group of related bacteria called "ORFan" genes are highly present in new operons. Many of the ORFans in operons are functional protein-encoding genes that contribute to the organism's fitness (Daubin and Ochman 2004). If more than one ORFan gene is grouped in an operon, it denotes that an unsequenced HGT source has introduced the ORFan genes. It is not a mandate for HGT to form an operon. Operons with native genes can form without HGT due to rearrangements of genes. After formation, many operons die since they are conserved across all or even most bacteria. Operons can also be lost if one/more genes are deleted or if an operon splits apart.

C. J. Betsy, C. Siva, *Fisheries Biotechnology and Bioinformatics*,
https://doi.org/10.1007/978-981-99-6991-3_5

5.2 Components of Operon

Picknett and Brenner (2001) described the following components of operon:

- Structural genes (S1–Sn): They form the primary structures of enzyme proteins that are involved in a metabolic pathway. E.g., biosynthesis of amino acid.
- Promoter (P): It is a short sequence of DNA which acts as the starting point to which RNA polymerase binds.
- Operator (O): It is found next to the promoter and comprises of a short segment of DNA. It binds a regulator protein that can either repress or activate transcription.

5.3 Structure of Operon

In DNA, region where clusters of related genes are present is called operons. It consists of promoter, operator, and multiple-related genes (Fig. 5.1). The operator is located either within the promoter or between the promoter and the genes. Transcription is initiated when RNA polymerase binds to the promoter. The location of the operator is important either to allow or to prevent transcription of the genes into mRNA.

5.4 Functions of Operon

Operon is involved in gene expression and polypeptides synthesis based on stimuli. *Escherichia coli* contains genes clustered into operons and regulons. The Lac operon is involved in lactose degradation, the Trp operon in tryptophan biosynthesis, and the His operon in histidine biosynthesis. These operons are turned on when the gene products are needed.

Operons can be under negative or positive control. Operon is under positive control when gene expression is stimulated by the presence of a regulatory protein. When operon is under negative control, it turns off the operon in the presence of a repressor which can be either repressible or inducible. Repressible operons are

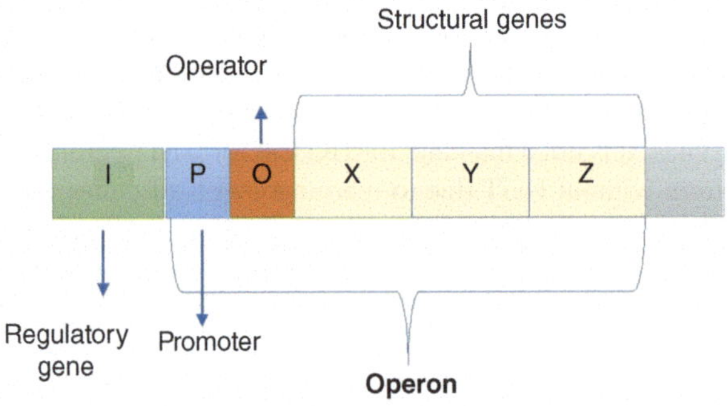

Fig. 5.1 Structure of an operon

involved in anabolic pathways while inducible operons are involved in catabolic pathways. Repressible operon is usually on but can be repressed in the presence of a repressor molecule. The repressor binds to the operator in order to block the RNA polymerase binding so that transcription doesn't occur. In contrast to repressible operon, inducible operon is usually off and its absence, the operator is blocked by a repressor molecule. In its presence, the inducer interacts with the repressor protein and releases it from the operator and allows transcription to proceed.

5.5 Lac Operon

It is the first operon to be identified by Jacob and Monod. The Lac operon is responsible for lactose degradation. Its expression is controlled by the regulatory gene LacI which encodes an allosteric repressor protein that keep the Lac operon "off." Repressor protein blocks the operator in the absence of lactose and hence is an inducible operon. Components of Lac operon (Fig. 5.2) are as follows:

- A promoter with operator
- Three genes such as
  - lacZ encode for β-galactosidase that breaks down lactose into glucose and galactose
  - lacY encode for permease involved in the metabolic process
  - lacA encode for transacetylase involved in the metabolic process

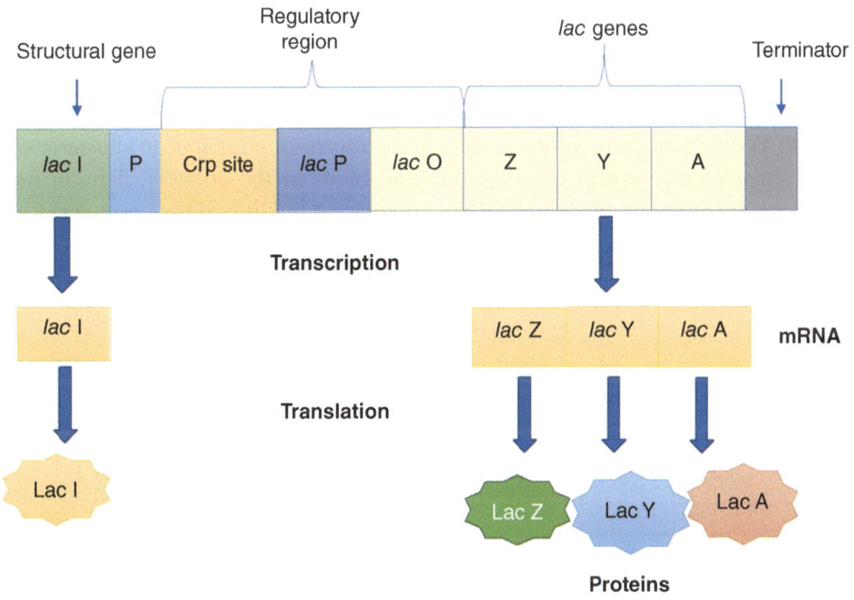

Fig. 5.2 Components of Lac operon

When an inducer molecule such as allolactose (isomer of lactose) inactivates the repressor protein, the Lac operon will be turned on. In the presence of lactose and its isomer, allolactose will bind to allosteric sites on the repressor protein which changes its conformation making it inactive. As the repressor protein detaches from the operator, RNA polymerase binds to the promoter leading to transcription and synthesis of three lactose degradation genes.

The gene regulation is positive in Lac operon. In order to synthesize the lacZ, lacY, and lacA genes, in the presence of lactose, repressor protein must be removed and during this, the gene expression will remain low. The amount of glucose in the cell decides the level of gene expression. This control is regulated by catabolite activator protein (CAP), an allosteric regulatory protein (Fig. 5.3). The level of glucose and the organic molecule cyclic AMP in the cell are inversely proportional to each other. Cyclic AMP activates CAP by binding to the allosteric sites, causing CAP to attach to the Lac operon promoter stimulating gene expression. This will increase the level of cell glucose and decrease the level of cyclic AMP and therefore, activator protein will disassociate from the promoter. Transcription will return to low levels or will turn off if the repressor protein reattaches.

5.6 Trp Operon

When the amino acid tryptophan is unavailable in the environment, it is synthesized by the Trp operon. The regulatory gene trpR regulates it. It is a repressible operon that will be turned on unless turned off by a repressor protein synthesized by trpR. The components of Trp operon (Fig. 5.4) include the following:

- Promoter with an operator
- Five genes

The repressor protein is synthesized in an inactive form and is always present in the cell. Tryptophan, a corepressor, binds to the repressor protein in an allosteric site which changes the protein conformation and binds to the operator. This in turn blocks transcription by preventing RNA polymerase binding to the promoter. This saves cell energy by not producing tryptophan when it is already present.

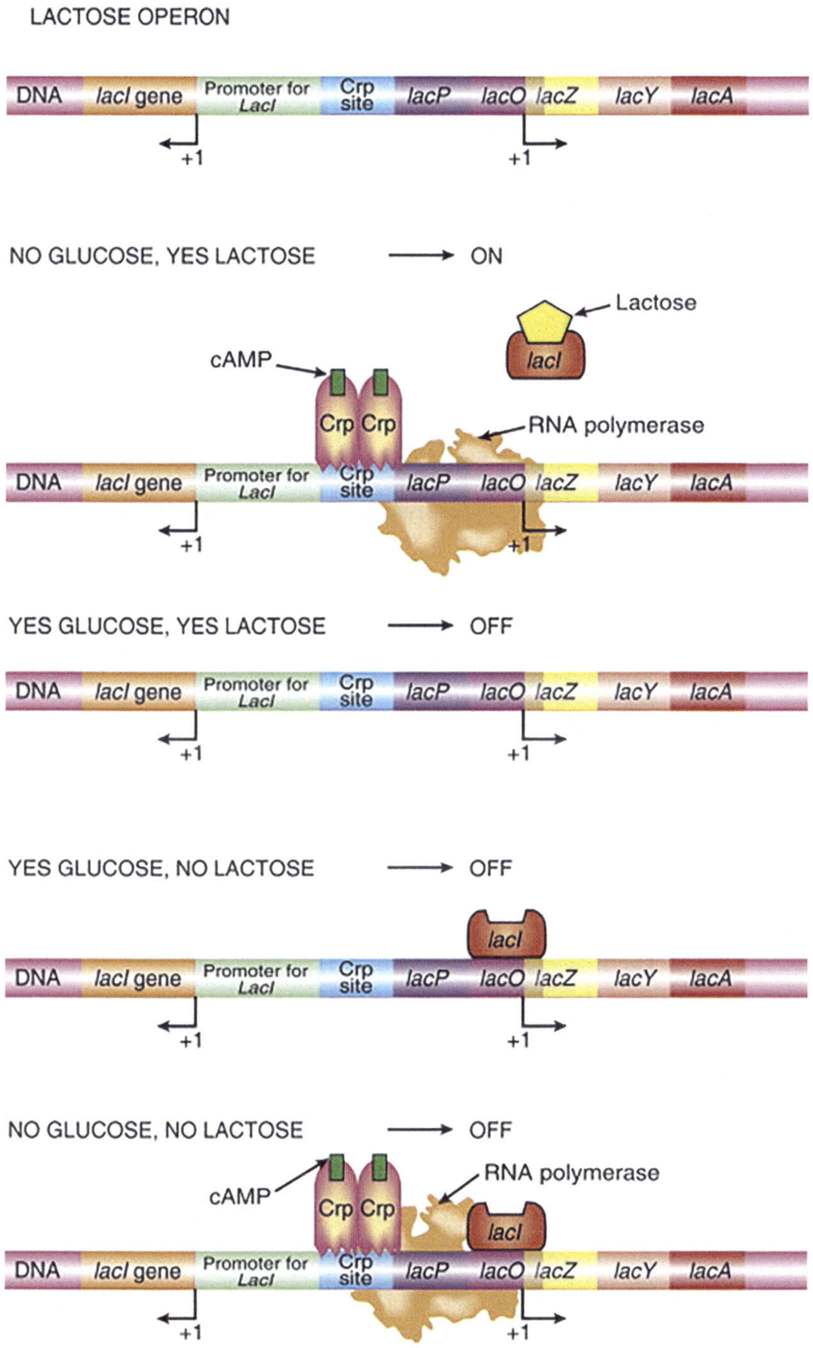

Fig. 5.3 The Lac operon and its control elements (reproduced from Clark and Pazdernik 2016 with permission from Elsevier)

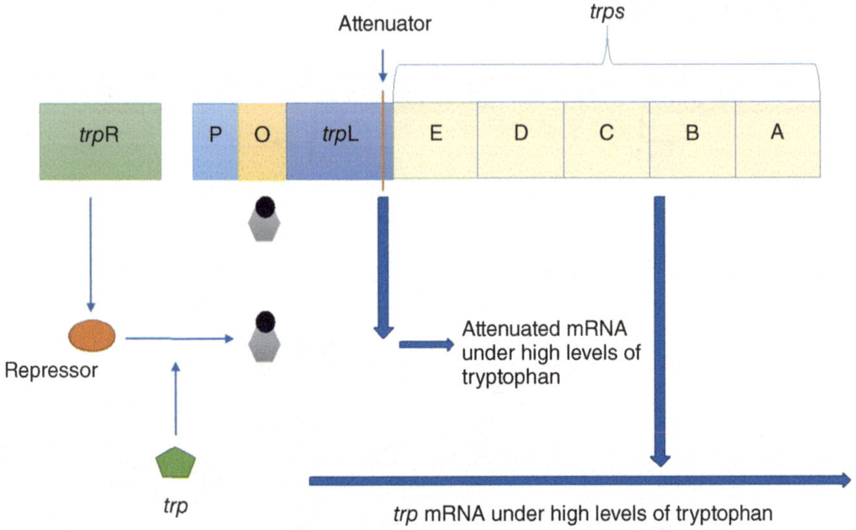

Fig. 5.4 Trp operon of *E. coli*

References

Clark DP, Pazdernik NJ (2016) DNA, RNA, and protein. In: Biotechnology, pp 33–61. https://doi.org/10.1016/B978-0-12-385015-7.00002-8

Daubin V, Ochman H (2004) Bacterial genomes as new gene homes: the genealogy of ORFans in E. coli. Genome Res 14:1036–1042

Picknett TM, Brenner S (2001) Operon. In: Encyclopedia of genetics, vol 1377. https://doi.org/10.1006/rwgn.2001.0936

Ramos JL, García-Salamanca A, Molina-Santiago C, Udaondo Z (2013) Operon. In: Brenner's encyclopedia of genetics, pp 176–180. https://doi.org/10.1016/b978-0-12-374984-0.01

Rocha EPC (2008) The organization of the bacterial genome. Annu Rev Genet 42:211–233

Gene Expression in Eukaryotes 6

Eukaryotes are multicellular organisms and thus have a complex process of gene regulation. In eukaryotes, the gene sequences are not continuous to produce protein. During mRNA production, the unwanted RNA regions are removed (introns), and the regions coding for amino acids (exons) are joined together by a process known as splicing. The gene expression mechanism in eukaryotes always ensures that each cell uses the same proteins it needs at any given moment.

6.1 DNA Replication, Transcription, and Translation

6.1.1 Replication

DNA must be duplicated before cell division so that each new cell contains the total amount of DNA material after cell division. Replication is the process of DNA duplication where the strands of the original double-stranded DNA molecule serve as a template for the complementary strand resulting in the production of two identical DNA molecules. The replicated new cell will have both original and newly synthesized DNA strand and hence this replication is termed as semi-conservative. In the genome, DNA replication begins at specific locations called "origins." Unwinding DNA at the region and synthesis of new strands form a replication fork.

Many enzymes are involved in the replication process. The enzyme helicase helps the double helix DNA to unwind in response to an initial synthesis of a short RNA strand. The unwound DNA strands are held in position by available proteins and serves as a template for replication. Primer, a small fragment of DNA or RNA is used to begin the synthesis of a new strand by pairing it with the template strand. DNA polymerase III joins the nucleotide units and extends an existing DNA strand. Once a primer pairs with DNA to be replicated, DNA polymerase produces new DNA strand by extending the 3′ end of an existing nucleotide chain by adding new

C. J. Betsy, C. Siva, *Fisheries Biotechnology and Bioinformatics*,
https://doi.org/10.1007/978-981-99-6991-3_6

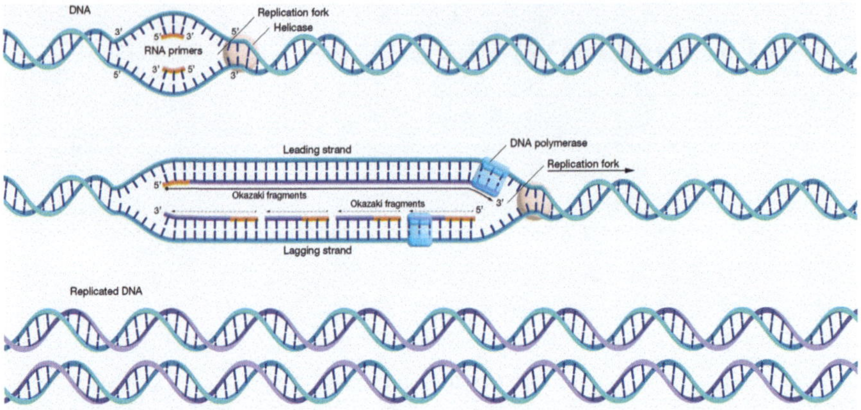

Fig. 6.1 DNA replication (reproduced with permission from Darryl Leja, National Human Genome Research Institute)

matching nucleotides to the template strand (Fig. 6.1). DNA polymerase is generally highly accurate and has proofreading ability to correct mismatched bases.

6.1.1.1 Origin of Replication

DNA must be replicated for a cell to divide. In DNA, this replication starts at specific points called "origins" which contains proteins that separate the two strands and initiate DNA synthesis. Origins contain DNA sequences recognized by replication initiator proteins (e.g., DnaA in *E. coli* and the Origin Recognition Complex in yeast) which recruit other proteins to separate the two strands at origin, forming a bubble and initiate replication forks. Origins are generally "AT-rich" since they have two hydrogen bonds which can be easily broken. Once strands are separated, RNA primers are created on the template strands.

6.1.1.2 Replication Fork

During replication, the hydrogen bonds at the origin holding the two DNA strands together are broken down by the enzyme helicase. This results in a structure that has two branching prongs with single DNA strand. These strands act as the template for the formation of leading and lagging strands with the help of DNA polymerase (Fig. 6.2).

6.1.1.3 Leading and Lagging Strands

The leading strand receives one RNA primer per active origin of replication, and the replication fork is directed towards the 3′ to 5′ direction which leads to the synthesize of complementary strand in the opposite direction. In case of prokaryotes, DNA polymerase III "reads" the DNA and adds nucleotides to the leading strand while in eukaryotes, it is Pol ε.

In contradictory to the leading strand, several RNA primers are received by the lagging strand and these multiple RNA primers present on the lagging strand are

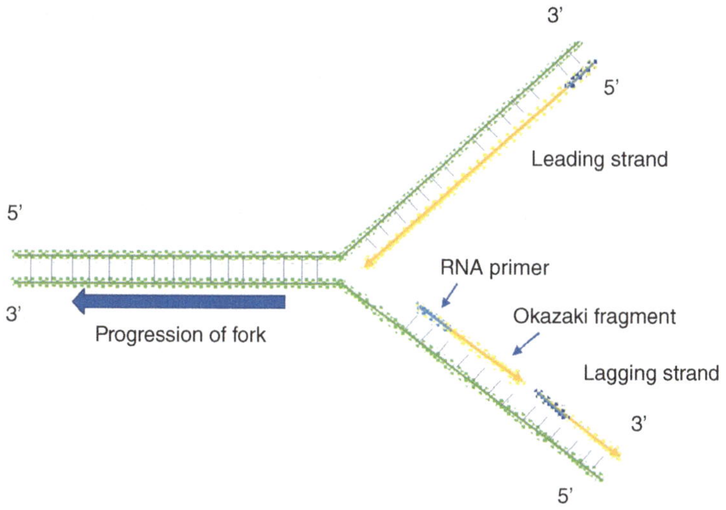

Fig. 6.2 Replication fork

called Okazaki fragments. In the lagging strand, the replication fork moves in a 5′ to 3′ direction which is opposite to the working orientation of DNA polymerase III and hence replication of the lagging strand is more complicated. On the lagging strand, primase "reads" the DNA and adds RNA in short separated segments.

DNA polymerase extends the leading strand in one continuous motion and the lagging strand in a discontinuous motion due to the presence of Okazaki fragments. RNAse removes the RNA fragments used to initiate replication by DNA polymerase, and another DNA polymerase enters to fill the gaps. When the process is complete, a single and several nicks can be found on the leading and lagging strands, respectively. These nicks are filled in by the enzyme ligase, and the newly replicated DNA molecule is completed.

6.1.1.4 Types of Replication

1. *Semi-conservative replication* would produce two copies containing original and new strands (Fig. 6.3). DNA replication in all cells follows this replication which was confirmed by the Meselson–Stahl experiment.
2. *Conservative replication* leaves the two original template DNA strands together in a double helix and produces a copy composed of two new strands containing all of the new DNA base pairs.
3. *Dispersive replication* produces two copies of the DNA, containing distinct regions of DNA composed of either original or new strands.

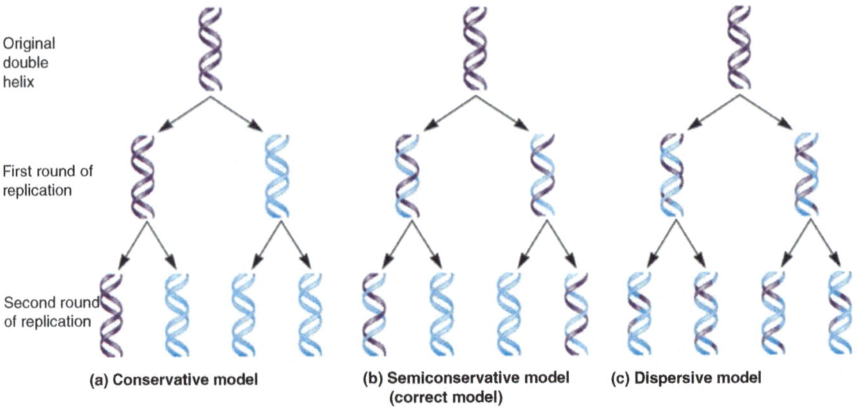

Fig. 6.3 Types of DNA replication (reproduced from Banerjee 2022 with permission from Springer Nature)

6.1.2 Transcription

Transcription occurs in the nucleus. During transcription, DNA template synthesizes mRNA and only one strand of DNA is copied. One gene can be copied many times. After transcription, the DNA strands rejoin. Transcription is carried out by the processes of initiation, elongation, and termination (Fig. 6.4).

6.1.2.1 Initiation

DNA contains a promoter region which identifies the start of a gene, decides the strand and direction in which it is to be copied. This promoter region is recognized, and RNA polymerase binds to it and unwinds the DNA. This disrupts the base pairs and produces a "bubble" of single-stranded DNA. Transcription occurs in a 5′ to 3′ direction, and the new ribonucleotide is added to the 3′ end. Since the binding between RNA polymerase and promoters occurs in a defined orientation, the given promoter always transcribes the same strand.

6.1.2.2 Elongation

The complementary bases of the copied DNA strand is assembled by the RNA polymerase. After synthesizing ten bases of RNA, it moves to the elongation phase. During elongation, the enzymes unwinds and re-anneals the DNA. As it moves along, the growing RNA chain is dissociated from the template and proofreading is also performed.

6.1.2.3 Termination

After transcribing the length of the gene, the polymerase will stop and release the RNA product by a process called termination. A termination code in the DNA indicates where transcription will stop. The mRNA produced is called an mRNA transcript.

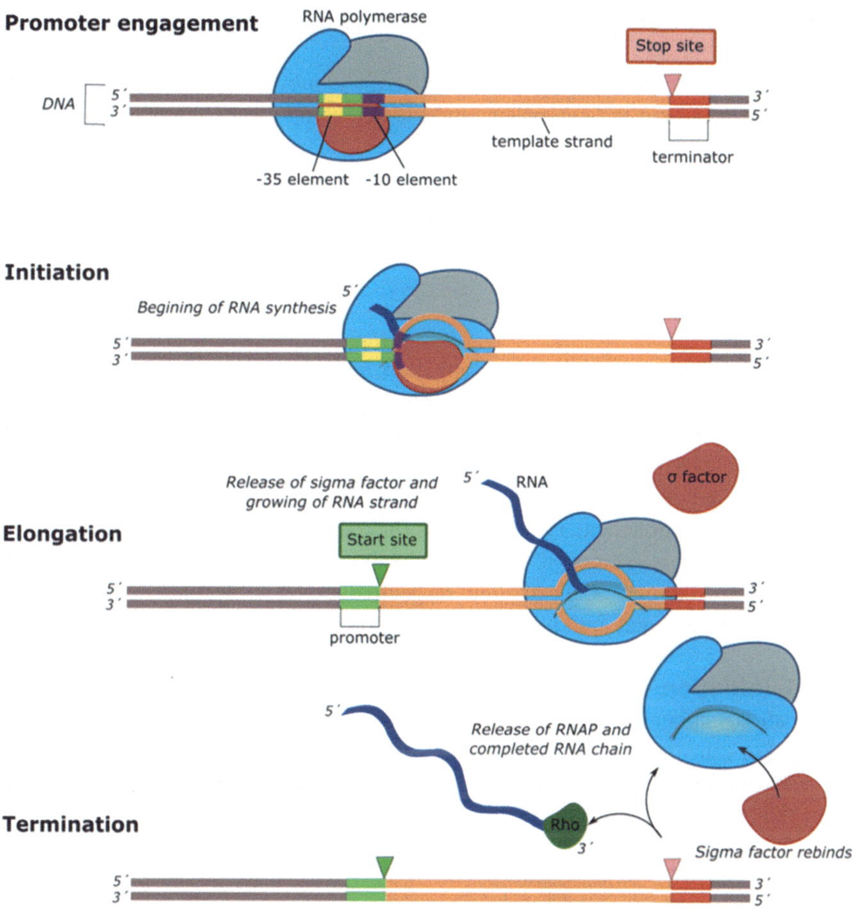

Fig. 6.4 Process of transcription (reproduced from Abril et al. 2020 with permission from Springer Nature)

6.1.2.4 Processing the mRNA Transcript

The mRNA transcript of eukaryotes contains the exons (coding sequences) and introns (intervening sequences) which must be processed for further use. They also have regions that do not contain a genetic message. These too short exons and too long introns are usually present in an alternating pattern and hence the genes bearing non-coding interruptions are said to be "split." A cap is added to the 5′ end, and a poly-A tail (150 to 200 Adenines) is added to the 3′ end of the molecule. The introns are removed from the primary transcripts of split genes through RNA splicing before they are translated into protein. This converts the pre-mRNA into mature messenger RNA which enters the cytoplasm from the nucleus through the pores in the nuclear membrane.

6.1.3 Translation

During translation, the mature mRNA transcript is used to synthesize proteins in ribosome. It occurs in the cytoplasm, and the key components required for translation are as follows:

- mRNA
- Ribosomes
- tRNA
- aminoacyl-tRNA synthetases.

Translation consists of initiation, elongation, and termination. During translation, mRNA nucleotide bases are read as three base codons, each of which codes for a particular amino acid. The genetic code is described as degenerate because a single amino acid may be coded by more than one codon. Specific codons also signal the start and the end of translation (Fig. 6.5).

Anticodons complementary to the mRNA codon is possessed by each tRNA molecule which is responsible for polypeptide assembly. Aminoacyl-tRNA synthetases link amino acids to their respective tRNA molecules. The resulting complex is charged and are known as aminoacyl-tRNA.

6.1.3.1 Initiation

The small 40s subunit of the ribosome with methionyl-tRNA binds at the 5′ cap of mRNA. For translation to start, the start codon 5′, AUG must be recognized, which is specific to methionine (anticodon on tRNA = 5′CAU). The large 60s subunit of

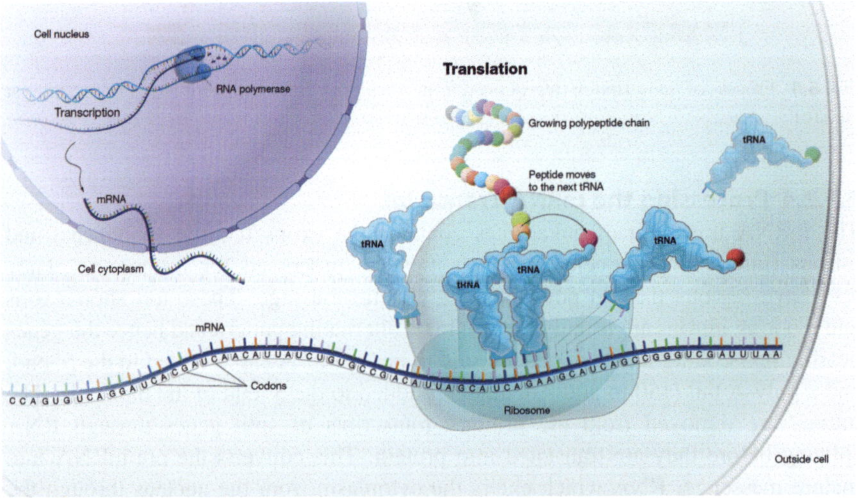

Fig. 6.5 Process of translation (reproduced with permission from Darryl Leja, National Human Genome Research Institute)

the ribosome then binds for elongation to occur. There are two tRNA binding sites in the ribosome; one is the P site which holds the peptide chain, and the other is the A site, which accepts the tRNA.

6.1.3.2 Elongation

During elongation, P site is occupied by Met-tRNA and A site is occupied by aminoacyl-tRNA which requires GTP. A peptide bond forms between methionine and the next aminoacyl-tRNA with the help of the enzyme peptidyl-transferase. The tRNA molecule in the P site leaves the ribosome since it becomes uncharged. Hence the ribosome translocates to the next codon along the mRNA molecule which opens up A site for the next aminoacyl-tRNA. The polypeptide chain is built up in the direction from the N terminal to the C terminal.

6.1.3.3 Termination

In termination, at A site, one of the three stop codons (UAG, UAA, and UGA) enters. tRNA molecules don't bind to these codons. This leads to hydrolyzation of the peptide and tRNA in the P site which releases the polypeptide into the cytoplasm. For the next translation process, the small and large subunits of the ribosome dissociate readily. The polypeptide produced undergoes post-translational modification before becoming a fully active protein (Fig. 6.6).

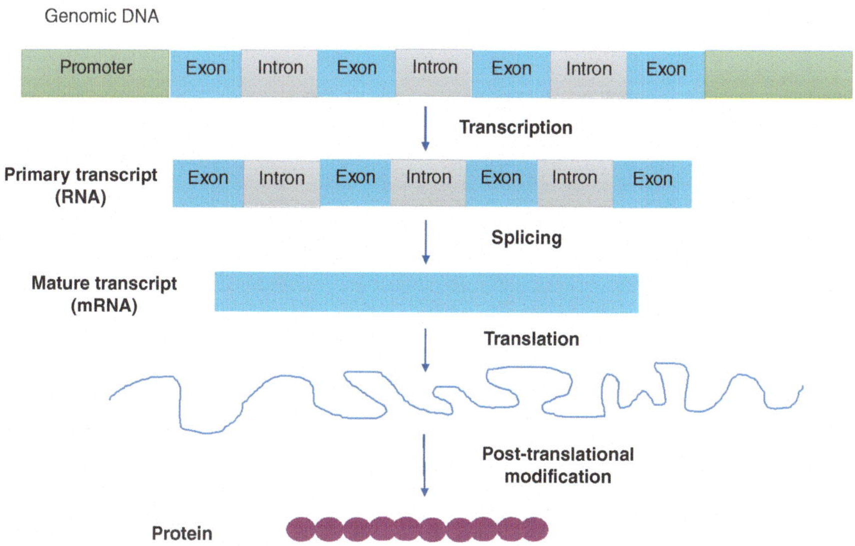

Fig. 6.6 Expression of gene through transcription and translation

References

Abril AG, Rama JLR, Sánchez-Pérez A et al (2020) Prokaryotic sigma factors and their transcriptional counterparts in *Archaea* and *Eukarya*. Appl Microbiol Biotechnol 104:4289–4302. https://doi.org/10.1007/s00253-020-10577-0

Banerjee T (2022) Replication of DNA. In: Kar D, Sarkar S (eds) Genetics fundamentals notes. Springer, Singapore. https://doi.org/10.1007/978-981-16-7041-1_7

Epigenetics

7

The term epigenetics was coined by Waddington in 1942 and was used to describe the "interactions of genes with their environment that brings the phenotype into being" (Waddington 2012). In general, genetic factors responsible for change in the appearance of an organism or biological functions without changing the actual DNA sequence is referred as epigenetics (Gavery and Roberts 2017). Therefore, only the expression of genes changes but not the genes. At molecular level, epigenetics refers to the addition or deletion of a methyl group to a DNA base, turning the gene on or off, or to packaging of the chromatin structure by silencing or opening regions of the genome by winding or unwinding the DNA around histones. Environmental factors are known to cause the epigenetic changes, and these changes are more common than mutations. It is also known that epigenetic changes caused by environmental exposures can be transmitted down several generations. The widely studied molecular epigenetic mechanisms are as follows:

- Regulation of chromatin structure through histone post-translational modifications
- Covalent modification of DNA principally through DNA methylation

However, other epigenetic molecular mechanisms such as regulation of gene expression through non-coding RNAs, and recombination of non-genic DNA, are also known to exist. Since environment plays a major role in determining the epigenetic mechanism, understanding the epigenetics is of great deal in aquaculture.

7.1 Covalent Modification of DNA: Cytosine Methylation

DNA methylation is a mechanism by which DNA can be epigenetically marked. DNA methylation is a modification of a cytosine C5 side-chain that adds a -CH_3 group through a covalent bond and is catalyzed by DNA methyltransferases

C. J. Betsy, C. Siva, *Fisheries Biotechnology and Bioinformatics*,
https://doi.org/10.1007/978-981-99-6991-3_7

Fig. 7.1 Mechanism of DNA methylation

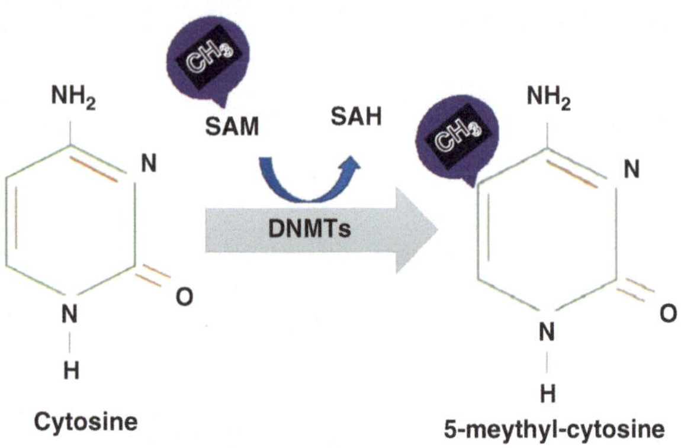

(DNMTs) (Okano et al. 1998). DNMTs transfer methyl groups to cytosine nucleotides within a continuous stretch of DNA, specifically at the 5-position of the pyrimidine ring (Fig. 7.1) (Santi et al. 1983; Chen et al. 1991). In order to be methylated, cytosines must be immediately followed by a guanine (Bird 1978; Cedar et al. 1979). Approximately 3% of the methylated cytosines are alone located at the 5′ end of genes, while the remaining 97% are found in intra- and intergenic sequences and within DNA repeats (Ballas et al. 2001). There are maintenance and de novo DNMTs. DNMT1 is the maintenance DNMT while DNMTs 3a and 3b are the de novo DNMT isoforms. These two DNMTs differ in the conditions under which they will methylate DNA.

7.2 Methylation Regulates Transcription Through Multiple Mechanisms

It has been reported that DNA methylation can be associated with transcriptional activation (Chahrour et al. 2008; Cohen et al. 2008). Apart from leading to structural changes in chromatin that can downregulate transcription significantly, DNA methylation can also directly interfere with the ability of transcription factors to bind to DNA regulatory elements at specific nucleotide sequences. For example, the transcription factor Ets-1 and the boundary element CCCTC binding factor (CTCF) can efficiently bind to non-methylated, but not methylated, DNA (Bell and Felsenfeld 2000).

7.3 DNA Methylation in Fishes

Studies have been carried to understand the patterns and functions of DNA methylation in zebrafish and medaka (Metzger and Schulte 2016). In general, vertebrates exhibit a global DNA methylation pattern, which means that most CpG dinucleotides (cytosine located at 5′ end of a guanine) are methylated with the exception of regions of DNA with a high CpG content, called as CpG islands. It is reported that global DNA methylation levels in fish are higher than those in mammals (Zhang et al. 2016). It is to be noted that the pattern of DNA methylation is "global" in vertebrates, whereas in case of invertebrates the pattern is "mosaic" (Simmen et al. 1999). In mammals, generally 60–70% of CpGs are methylated, whereas in case of Pacific oyster, only 15% of CpGs were methylated when the whole genome was analyzed (Gavery and Roberts 2013). This methylated fraction in oysters consists of gene bodies, and the transposable elements in oysters are less methylated (Feng et al. 2010; Zemach et al. 2010). However, DNA methylation is reported to be involved in shellfish gene regulation. Recently, it was stated that parental exposure of an oyster to a herbicide influences progeny DNA methylation patterns (Rondon et al. 2017).

7.4 Histone Modifications

The function of histones is organizing DNA within the nucleus. In the nucleus, DNA is tightly packaged into chromatin, a DNA–protein complex that consists of DNA in a double helix, histone proteins, and many associated regulatory proteins. Modification of histones is a crucial mechanism for epigenetic tagging of the genome (Strahl and Allis 2000). Histone modification can occur as a consequence of DNA methylation or can be controlled by intracellular signaling. The basic unit of chromatin is the nucleosome, which is composed of an octamer of histone proteins (containing two copies each of histones 2A, 2B, 3, and 4) around which is wrapped—like a rope on a windlass—the DNA double helix (Fig. 7.2). The degree to which nucleosomes are condensed or packed is a critical determinant of the transcriptional activity of the associated DNA and this is mediated in part by chemical modifications of the N-terminal tails of histone proteins. Structural studies indicate that these N-terminal tails protrude from the nucleosome and are extensively modified post-translationally (Luger et al. 1997).

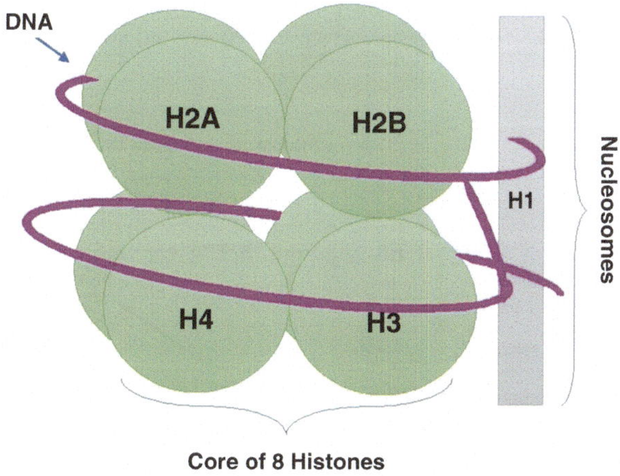

Fig. 7.2 Histone octamers

The major post-translational modifications of histone tails include the following:

- Acetylation
- Methylation
- Ubiquitination
- Phosphorylation

7.4.1 Acetylation

- Occurs on ε-NH+ group of the lysine residues.
- Mediated by histone acetyltransferases.
- Acetylation of lysine neutralizes the positive charge.
- During acetylation, acetyl group is added by acetyl transferase on lysine which leads to the adoption of a relaxed state of chromatin called "euchromatin" (Fig. 7.3). In this state, transcription factors and other proteins can bind to their DNA binding sites and proceed with active transcription (Kim and Kaang 2017).
- It is a reversible process, and the enzymes that catalyze the reversal of histone acetylation are known as HDACs.
- Classical isoforms of HDACs catalyze the removal of acetyl groups from lysine residues through a Zn2+-dependent charge-relay system.

7.4.2 Methylation

- Occurs on ε-NH+ group of the lysine residues.
- Mediated by lysine methyltransferases.
- Methylation of lysines preserves their positive charge.

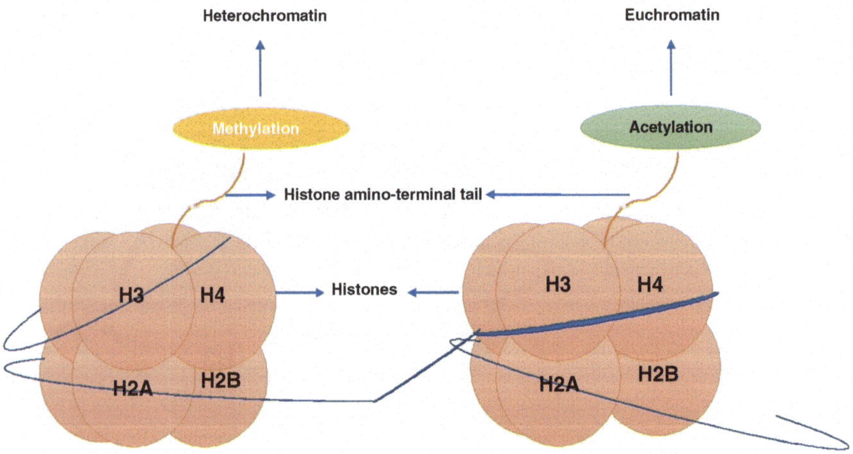

Fig. 7.3 Histone methylation and acetylation

- During methylation, methyl is added by histone methyltransferase on lysine and arginine which in turn leads to the adoption of a condensed state of chromatin called "heterochromatin." (Fig. 7.3)
- Heterochromatin blocks transcription machinery from binding to DNA and results in transcriptional repression (Kim and Kaang 2017).
- Lysines can accept up to three methyl groups (mono-, di-, or trimethylated states).
- Arginine residues within histones can also be mono- or dimethylated on their guanidine nitrogen which is catalyzed by protein arginine methyltransferases.

7.4.3 Ubiquitination

- Ubiquitin is a protein with 76 amino acids that has ubiquitous distribution in all cell types and high degree of conservation across species (Pickart 2001).
- Ubiquitin is an 8.5-kDa polypeptide with a flexible six-amino-acid carboxy-terminal tail terminating with a glycine residue (Jeusset and McManus 2019).
- When a ubiquitin is attached to the ε-NH+ group of a lysine, ubiquitination of histones occur (Nickel and Davie 1989).
- Histones can either be mono-ubiquitinated or poly-ubiquitinated.
- Three enzymes are involved in the addition of a ubiquitin moiety to a target protein.
- Ubiquitin is first loaded onto a ubiquitin-activating enzyme (E1), in an ATP-dependent reaction, before being transferred to a ubiquitin-conjugating enzyme (E2), and finally ligated to the target lysine residue by a ubiquitin ligase (E3) which is responsible for substrate specificity (Fig. 7.4) (Buetow and Huang 2016; Groen and Gillingwater 2015).
- Protein ubiquitination levels are dynamically regulated by the opposing activity of deubiquitinating enzymes (DUBs), also known as ubiquitin-specific peptidases (USPs).

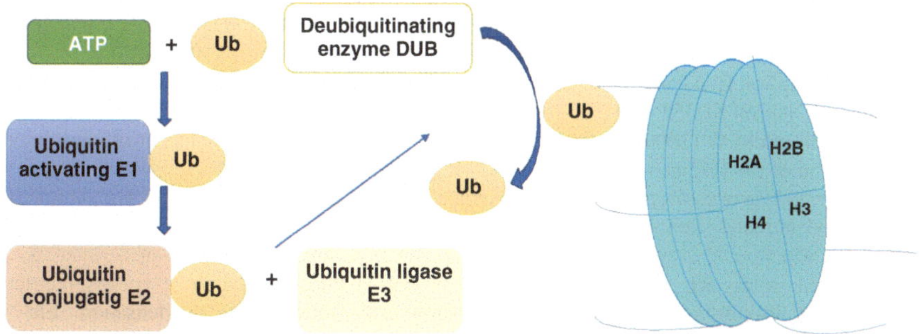

Fig. 7.4 Enzymatic mechanism of the addition and removal of histone ubiquitination

7.4.4 Phosphorylation

- It is associated with gene activation.
- It has been well studied in humans.
- During mitosis, histones H1 and H3 are reported to be involved in chromosome condensation (Bradbury et al. 1973; Gurley et al. 1974).
- H3 was the first histone whose phosphorylation was characterized in response to activation of mitogenic signaling pathways (Mahadevan et al. 1991).
- During phosphorylation, phosphate group is added by kinase on serine/tyrosine.
- Phosphatases remove phosphate groups from histones (Ajiro et al. 1996).
- To date, the phosphatases PP1 and PP2A, and the PP1 inhibitor DARPP32, have been shown to regulate H3 phosphorylation in the central nervous system (Hsu et al. 2000; Nowak et al. 2003; Stipanovich et al. 2008).

7.4.5 Histone Post-translational Modifications in Fish

Histone acetylation is involved in zebrafish embryogenesis (Vastenhouw and Schier 2012) and in tissue regeneration (Stewart et al. 2009). It was reported that diet influences bulk histone modification levels and can regulate the expression of associated enzymes in rainbow trout and European sea bass (Marandel et al. 2016; Panserat et al. 2017). Similar to mammals, zebrafish shows multivalent modified histone retention in sperm (Wu et al. 2011). The post-translational modification of the histone variant H2A.X was reported in Eastern oysters in response to algal toxins (Gonzalez-Romero et al. 2017). In Pacific oysters, homologs of Jumonji histone demethylase genes were identified which were regulated during embryonic development (Fellous et al. 2014). In oysters, bulk histone methylation levels and the expression of histone demethylases responded to temperature during development (Fellous et al. 2015).

References

Ajiro K, Yoda K, Utsumi K, Nishikawa Y (1996) Alteration of cell cycle-dependent histone phosphorylations by okadaic acid. Induction of mitosis-specific H3 phosphorylation and chromatin condensation in mammalian interphase cells. J Biol Chem 271:13197–13201

Ballas N, Battaglioli E, Atouf F et al (2001) Regulation of neuronal traits by a novel transcriptional complex. Neuron 31:353–365

Bell AC, Felsenfeld G (2000) Methylation of a CTCF-dependent boundary controls imprinted expression of the Igf2 gene. Nature 405:482–485

Bird AP (1978) Use of restriction enzymes to study eukaryotic DNA methylation: II. The symmetry of methylated sites supports semi-conservative copying of the methylation pattern. J Mol Biol 118:49–60

Bradbury EM, Inglis RJ, Matthews HR, Sarner N (1973) Phosphorylation of very-lysine-rich histone in Physarum polycephalum. Correlation with chromosome condensation. Eur J Biochem 33:131–139

Buetow L, Huang DT (2016) Structural insights into the catalysis and regulation of E3 ubiquitin ligases. Nat Rev Mol Cell Biol 17:626–642

Cedar H, Solage A, Glaser G, Razin A (1979) Direct detection of methylated cytosine in DNA by use of the restriction enzyme MspI. Nucleic Acids Res 6:2125–2132

Chahrour M, Jung SY, Shaw C et al (2008) MeCP2, a key contributor to neurological disease, activates and represses transcription. Science 320:1224–1229

Chen L, MacMillan AM, Chang W, Ezaz-Nikpay K, Lane WS, Verdine GL (1991) Direct identification of the active-site nucleophile in a DNA (cytosine-5)-methyltransferase. Biochemistry 30(46):11018–11025

Cohen S, Zhou Z, Greenberg ME (2008) Activating a repressor. Science 320:1172–1173

Fellous A, Favrel P, Guo X, Riviere G (2014) The Jumonji gene family in Crassostrea gigas suggests evolutionary conservation of Jmj-C histone demethylases orthologues in the oyster gametogenesis and development. Gene 538:164–175. https://doi.org/10.1016/j.gene.2013.12.016

Fellous A, Favrel P, Riviere G (2015) Temperature influences histone methylation and mRNA expression of the Jmj-C histone-demethylase orthologues during the early development of the oyster Crassostrea gigas. Mar Genomics 19:23–30. https://doi.org/10.1016/j.margen.2014.09.002

Feng S, Cokus SJ, Zhang X, Chen P-Y, Bostick M, Goll MG, Hetzel J, Jain J, Strauss SH, Halpern ME, Ukomadu C, Sadler KC, Pradhan S, Pellegrini M, Jacobsen SE (2010) Conservation and divergence of methylation patterning in plants and animals. Proc Natl Acad Sci U S A 107:8689–8694. https://doi.org/10.1073/pnas.1002720107

Gavery and Roberts (2017) Epigenetic considerations in aquaculture. PeerJ 5:e4147. https://doi.org/10.7717/peerj.4147

Gavery MR, Roberts SB (2013) Predominant intragenic methylation is associated with gene expression characteristics in a bivalve mollusc. PeerJ 1:e215. https://doi.org/10.7717/peerj.215

Gonzalez-Romero R, Suarez-Ulloa V, Rodriguez-Casariego J, Garcia-Souto D, Diaz G, Smith A, Pasantes JJ, Rand G, Eirin-Lopez JM (2017) Effects of Florida red tides on histone variant expression and DNA methylation in the eastern oyster Crassostrea virginica. Aquat Toxicol 186:196–204. https://doi.org/10.1016/j.aquatox.2017.03.006

Groen EJN, Gillingwater TH (2015) UBA1: at the crossroads of ubiquitin homeostasis and neurodegeneration. Trends Mol Med 21:622–632

Gurley LR, Walters RA, Tobey RA (1974) Cell cycle-specific changes in histone phosphorylation associated with cell proliferation and chromosome condensation. J Cell Biol 60:356–364

Hsu JY, Sun ZW, Li X et al (2000) Mitotic phosphorylation of histone H3 is governed by Ipl1/aurora kinase and Glc7/PP1 phosphatase in budding yeast and nematodes. Cell 102:279–291

Jeusset LM-P, McManus KJ (2019) Developing targeted therapies that exploit aberrant histone ubiquitination in cancer. Cell 8(2):165. https://doi.org/10.3390/cells8020165

Kim S, Kaang BK (2017) Epigenetic regulation and chromatin remodeling in learning and memory. Exp Mol Med 49:e281. https://doi.org/10.1038/emm.2016.140

Luger K, Mader AW, Richmond RK, Sargent DF, Richmond TJ (1997) Crystal structure of the nucleosome core particle at 2.8 a resolution. Nature 389:251–260

Mahadevan LC, Willis AC, Barratt MJ (1991) Rapid histone H3 phosphorylation in response to growth factors, phorbol esters, okadaic acid, and protein synthesis inhibitors. Cell 65:775–783

Marandel L, Lepais O, Arbenoits E, Véron V, Dias K, Zion M, Panserat S (2016) Remodelling of the hepatic epigenetic landscape of glucose-intolerant rainbow trout (Oncorhynchus mykiss) by nutritional status and dietary carbohydrates. Sci Rep 6:32187. https://doi.org/10.1038/srep32187

Metzger DCH, Schulte PM (2016) Epigenomics in marine fishes. Mar Genomics 30:43–54. https://doi.org/10.1016/j.margen.2016.01.004

Nickel BE, Davie JR (1989) Structure of polyubiquitinated histone H2A. Biochemistry 28:964–968

Nowak SJ, Pai CY, Corces VG (2003) Protein phosphatase 2A activity affects histone H3 phosphorylation and transcription in Drosophila melanogaster. Mol Cell Biol 23:6129–6138

Okano M, Xie S, Li E (1998) Cloning and characterization of a family of novel mammalian DNA (cytosine-5) methyltransferases. Nat Genet 19:219–220

Panserat S, Marandel L, Geurden I, Veron V, Dias K, Plagnes-Juan E, Pegourié G, Arbenoits E, Santigosa E, Weber G, Verlhac TV (2017) Muscle catabolic capacities and global hepatic epigenome are modified in juvenile rainbow trout fed different vitamin levels at first feeding. Aquaculture 468(Part 1):515–523. https://doi.org/10.1016/j.aquaculture.2016.11.021

Pickart CM (2001) Mechanisms underlying ubiquitination. Annu Rev Biochem 70:503–533

Rondon R, Grunau C, Fallet M, Charlemagne N, Sussarellu R, Chaparro C, Montagnani C, Mitta G, Bachère E, Akcha F, Cosseau C (2017) Effects of a parental exposure to diuron on Pacific oyster spat methylome. 3: article dvx004. Environ Epigenet 3(1):dvx004. https://doi.org/10.1093/eep/dvx004

Santi DV, Garrett CE, Barr PJ (1983) On the mechanism of inhibition of DNA-cytosine methyltransferases by cytosine analogs. Cell 33:9–10

Simmen MW, Leitgeb S, Charlton J, Jones SJ, Harris BR, Clark VH, Bird A (1999) Nonmethylated transposable elements and methylated genes in a chordate genome. Science 283:1164–1167. https://doi.org/10.1126/science.283.5405.1164

Stewart S, Tsun Z-Y, Izpisu Belmonte JC (2009) A histone demethylase is necessary for regeneration in zebrafish. Proc Natl Acad Sci U S A 106:19889–19894. https://doi.org/10.1073/pnas.0904132106

Stipanovich A, Valjent E, Matamales M et al (2008) A phosphatase cascade by which rewarding stimuli control nucleosomal response. Nature 453(7197):879–884

Strahl BD, Allis CD (2000) The language of covalent histone modifications. Nature 403:41–45

Vastenhouw NL, Schier AF (2012) Bivalent histone modifications in early embryogenesis. Curr Opin Cell Biol 24:374–386. https://doi.org/10.1016/j.ceb.2012.03.009

Waddington CH (2012) The epigenotype. Int J Epidemiol 41:10–13

Wu S-F, Zhang H, Cairns BR (2011) Genes for embryo development are packaged in blocks of multivalent chromatin in zebrafish sperm. Genome Res 21:578–589. https://doi.org/10.1101/gr.113167.110

Zemach A, McDaniel IE, Silva P, Zilberman D (2010) Genome-wide evolutionary analysis of eukaryotic DNA methylation. Science 328:916–919. https://doi.org/10.1126/science.1186366

Zhang C, Hoshida Y, Sadler KC (2016) Comparative epigenomic profiling of the DNA methylome in mouse and zebrafish uncovers high interspecies divergence. Front Genet 7:110. https://doi.org/10.3389/fgene.2016.00110

8 RNA in Gene Regulation

According to the central dogma of molecular biology, DNA stores genetic information and RNA transmits the genetic information. RNA plays an important role in controlling the gene expression and gene regulation. RNA can make RNA–RNA, RNA–DNA complexes which can control the expression of genes. Recently, many structural and functional diversity of RNA molecules has been discovered. These are noncoding RNAs (ncRNAs) which offer powerful gene silencing pathways. In humans, it is reported that only ≤2% of the total DNA are translated into proteins, and the remaining genes are transcribed into noncoding RNAs (Qi and Du 2013) and >98% of RNAs formed in human cells are ncRNAs (Rusk 2015). Antisense RNA (asRNA) represents a type of ncRNA, and it includes small interfering RNA (siRNA), microRNA (miRNA), long noncoding RNA (lncRNA), and PIWI interacting RNA (piRNA).

8.1 Antisense RNA

Antisense RNAs are also referred to as natural regulatory RNAs (Saberi et al. 2016). It is a single-stranded RNA that is complementary to a messenger RNA (mRNA) strand transcribed within a cell (Wittwer and Makrigiorgos 2018). They are small and diffusible molecules, containing 19–23 nucleotides that complement mRNA (Xu et al. 2018). They recognize sequence and structural elements that are present in themselves and their target mRNAs and carry out their regulatory function by inhibiting mRNA transcription and/or translation (Saberi et al. 2016). Sometimes, they activate expression of mRNAs (Wagner and Flardh 2002; Lioliou et al. 2010). mRNA translation is inhibited by binding to antisense RNA which is complementary to mRNA which may cause mRNA degradation and is known as post-transcriptional gene silencing.

In bacteria, asRNAs are divided into *cis*-acting 5′ element non-coding asRNAs, *trans*-acting small non-coding asRNAs, and *cis*-encoded asRNAs. A *cis*-acting 5′

C. J. Betsy, C. Siva, *Fisheries Biotechnology and Bioinformatics*,
https://doi.org/10.1007/978-981-99-6991-3_8

Table 8.1 Difference between *cis*- and *trans*-asRNAs (modified from Saberi et al. 2016)

Cis-asRNAs	*Trans*-asRNAs
Transcribed from the opposing DNA strand of the target gene at the same genomic locus	Transcribed from a separate locus to the target gene
Have perfect complementarity with the target gene	Have imperfect complementarity with the target gene
Target individual mRNAs	Target multiple sense targets and form complex regulation networks
Usually short	Usually long

non-coding asRNA is usually attached to the 5′ side of an mRNA, the expression of which is regulated by the non-coding RNA. *Trans*-acting small non-coding asRNAs are usually encoded in intergenic regions on the chromosome and control translation or degradation of their target mRNAs. *Cis*-acting asRNA is expressed as a complementary sequence of mRNA that becomes the sole target RNA (Saberi et al. 2016). The difference between *cis*- and *trans*- asRNAs is given in Table 8.1.

8.1.1 Small Interfering RNA

Double-stranded RNAs (dsRNAs) selectively destruct mRNAs which have the same sequence as the ds RNAs. This leads to selective inhibition of gene expression and is called RNA interference (RNAi). In cell, the dsRNA is first cleaved by a ribonuclease called Dicer, into small ds fragments of 21–23 nucleotides known as small interfering RNAs (Si RNAs). One strand of SiRNA is destroyed, and its antisense strand is taken up by an enzyme called RNA-induced silencing complex (RISC) to destroy the corresponding mRNA. This blocks the synthesis of proteins and inhibits the translation of mRNA or they can move into the nucleus where they cause chromatin modification.

In *Caenorhabditis elegans*, dsRNA is transferred to the cytoplasm by the transmembrane protein SID-1, and then processed by DCR-1 to form the ds siRNA. The ds siRNAs are connected to siRISC by RISC, and then are unfolded by a helicase to activate the siRISC which is associated with the target mRNA in a sequence-specific manner. RISC generated by a single siRNA can degrade the complementary target mRNA specifically and completely, thereby inhibiting the translation process (Saurabh et al. 2014).

8.1.2 Micro RNAs

miRNAs are single-stranded RNA (ssRNA) that contain 18–25 nucleotides and are found in eukaryotes (Xu et al. 2018). They break down their complementary mRNA using enzymes which prevents mRNA from being translated, or accelerate its degradation. It is completely different from the primary transcripts of miRNAs which are long RNA transcripts of noncoding DNA (Mohr and Mott 2015). miRNA cannot be translated into a protein but is used as an inhibitor of the expression of its target coding gene (Mohr and Mott 2015). The primary transcripts of miRNA contain hairpin-like miRNA precursor (Adams 2017) which is processed in the nucleus by enzyme ribonuclease (RNase) III to generate precursor miRNA (pre-miRNA). This intranuclear pre-miRNA is transferred from the nucleus to the cytoplasm by Exportin-5 which forms a novel pre-miRNA containing stem and loop structures (Ling et al. 2013). The pre-miRNAs are then cleaved at the hairpin stem region by RNase III (i.e., Dicer) to generate mature miRNAs (Kim et al. 2016). The mature miRNAs can be connected by the Argonaute protein family with RNA-induced silencing complex (RISC) to activate RISC (Riley et al. 2012), thus leading to the degradation of the target mRNA or the repression of translation (Fig. 8.1) (Nishimura and Fabian 2016). At this point, miRNAs regulate gene expression by the base complementarity between mRNAs and miRNAs rather than by mRNA degradation (Schmiedel et al. 2015).

8.1.3 lncRNAs

lncRNA is an mRNA-like transcript that ranges from 200 bp to 1000 kb in length and is not involved in protein synthesis (Legeai and Derrien 2015). Based on their position in the genome, they are divided into sense, antisense, bidirectional, intronic, and intergenic (Mattick 2009). It is transcribed by RNA polymerase II and then undergoes co-transcriptional modifications and finally, it is folded to form secondary and/or tertiary structures (Legeai and Derrien 2015).

These lncRNAs combine with gene transcripts to form a complementary ds structure, which can be processed by the enzyme Dicer to produce endogenous siRNAs. Furthermore, the unpaired lncRNA sequences can also complement mature miRNA molecules. Antisense lncRNAs are involved in X-chromosome inactivation, genomic imprinting, and the occurrence and development of certain diseases (Villegas and Zaphiropoulos 2015; Khani et al. 2016).

8.1.4 piRNAs

piRNAs contain 26–32 nucleotides and were first identified in mammalian testes (Zuo et al. 2016). Along with the members of PIWI protein family, piRNAs play a major role in regulating gene expression (Ross et al. 2014) and regulating the growth and development of germ cells (Nishida et al. 2015). They are mostly known to exist in mammalian germ cells and stem cells (Mourier 2011). It is known to be involved in silencing the gene-transcription process, maintaining the functions of stem cells and the reproductive system, and regulating mRNA stability and gene translation (Ross et al. 2014).

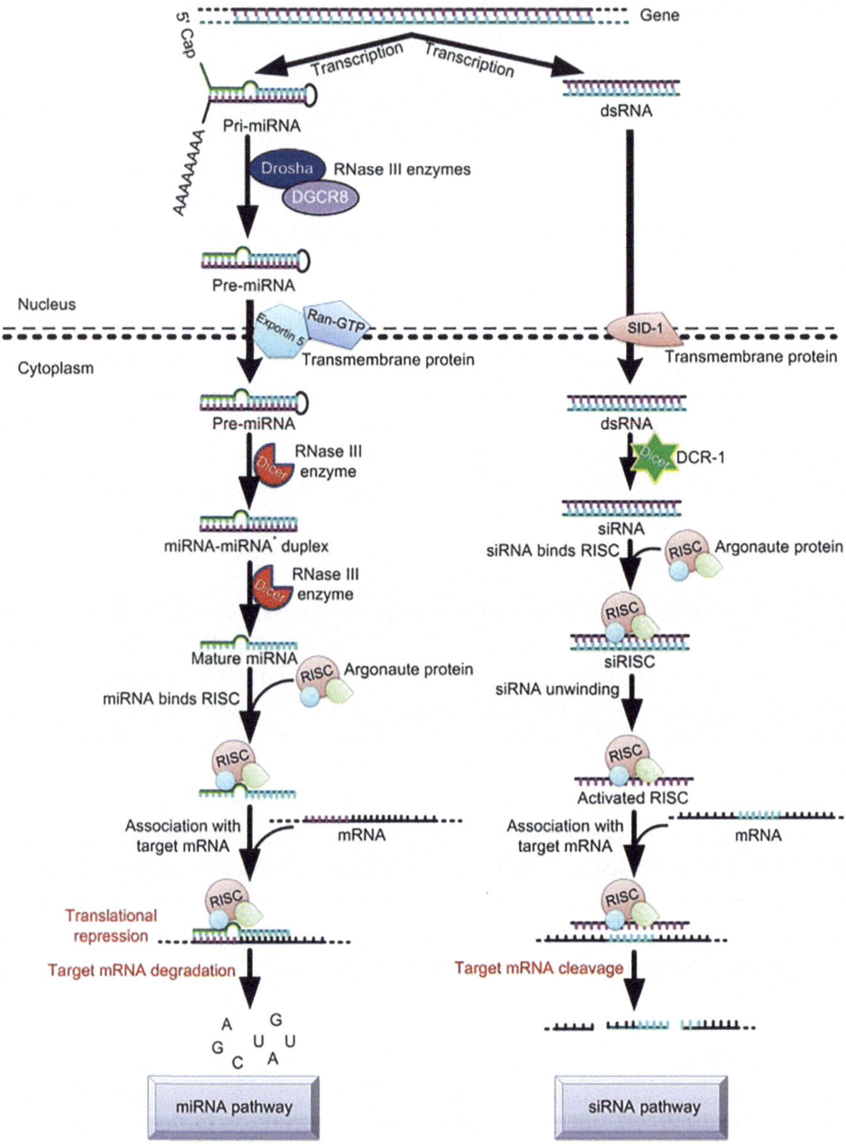

Fig. 8.1 Formation and regulatory mechanisms of miRNAs and siRNA (reproduced from Xu et al. 2018 with permission from Springer Nature)

8.1.5 Ribozymes

Tom Cech and Sidney Altman discovered that RNA could act as an enzyme. Ribozymes are antisense RNA molecules with enzymatic properties which bind to the target RNA by cleaving phosphodiester backbone at a specific site. They have two binding sites such as a substrate binding site and a guanosine binding site.

References

Adams L (2017) Non-coding RNA: pri-miRNA processing: structure is key. Nat Rev Genet 18(3): 145. https://doi.org/10.1038/nrg.2017.6

Khani MH, Yeganeh F, Sotoodehnejadnematalahi F (2016) Long non-coding RNAs; new perspective for autoimmune disease. MOJ Immunol 3(3):00090. https://doi.org/10.15406/moji.2016.03.00090

Kim YK, Kim B, Kim VN (2016) Re-evaluation of the roles of DROSHA, Exportin 5, and DICER in microRNA biogenesis. Proc Natl Acad Sci U S A 113(13):E1881–E1889. https://doi.org/10.1073/pnas.1602532113

Legeai F, Derrien T (2015) Identification of long non-coding RNAs in insects genomes. Curr Opin Insect Sci 7:37–44. https://doi.org/10.1016/j.cois.2015.01.003

Ling H, Fabbri M, Calin GA (2013) MicroRNAs and other non-coding RNAs as targets for anticancer drug development. Nat Rev Drug Discov 12(11):847–865. https://doi.org/10.1038/nrd4140

Lioliou E, Romilly C, Romby P, Fechter P (2010) RNA-mediated regulation in bacteria: from natural to artificial systems. Nat Biotechnol 27:222–235

Mattick JS (2009) The genetic signatures of noncoding RNAs. PLoS Genet 5(4):e1000459. https://doi.org/10.1371/journal.pgen.1000459

Mohr AM, Mott JL (2015) Overview of microRNA biology. Semin Liver Dis 35(1):3–11. https://doi.org/10.1055/s-0034-1397344

Mourier T (2011) Retrotransposon-centered analysis of piRNA targeting shows a shift from active to passive retrotransposon transcription in developing mouse testes. BMC Genomics 12:440. https://doi.org/10.1186/1471-2164-12-440

Nishida KM, Iwasaki YW, Murota Y et al (2015) Respective functions of two distinct Siwi complexes assembled during PIWI-interacting RNA biogenesis in Bombyx germ cells. Cell Rep 10(2):193–203. https://doi.org/10.1016/j.celrep.2014.12.013

Nishimura T, Fabian MR (2016) Scanning for a unified model for translational repression by microRNAs. EMBO J 35(11):1158–1159. https://doi.org/10.15252/embj.201694324

Qi P, Du X (2013) The long non-coding RNAs, a new cancer diagnostic and therapeutic gold mine. Mod Pathol 26(2):155–165. https://doi.org/10.1038/modpathol.2012.160

Riley KJ, Yario TA, Steitz JA (2012) Association of Argonaute proteins and microRNAs can occur after cell lysis. RNA 18(9):1581–1585. https://doi.org/10.1261/rna.034934.112

Ross RJ, Weiner MM, Lin HF (2014) PIWI proteins and PIWI-interacting RNAs in the soma. Nature 505(7483):353–359. https://doi.org/10.1038/nature12987

Rusk N (2015) Understanding noncoding RNAs. Nat Methods 12(1):35. https://doi.org/10.1038/nmeth.3235

Saberi F, Kamali M, Najafi A, Yazdanparast A, Moghaddam MM (2016) Natural antisense RNAs as mRNA regulatory elements in bacteria: a review on function and applications. Cell Mol Biol Lett 21(1):6. https://doi.org/10.1186/s11658-016-0007-z

Saurabh S, Vidyarthi AS, Prasad D (2014) RNA interference: concept to reality in crop improvement. Planta 239(3):543–564. https://doi.org/10.1007/s00425-013-2019-5

Schmiedel JM, Klemm SL, Zheng YN et al (2015) MicroRNA control of protein expression noise. Science 348(6230):128–132. https://doi.org/10.1126/science.aaa1738

Villegas VE, Zaphiropoulos PG (2015) Neighboring gene regulation by antisense long non-coding RNAs. Int J Mol Sci 16(2):3251–3266. https://doi.org/10.3390/ijms16023251

Wagner E, Flardh K (2002) Antisense RNAs everywhere? Trends Genet 18:223–226

Wittwer CT, Makrigiorgos GM (2018) Nucleic acid techniques. In: Principles and applications of molecular diagnostics, pp 47–86. https://doi.org/10.1016/B978-0-12-816061-9.00004-7

Xu JZ, Zhang JL, Zhang WG (2018) Antisense RNA: the new favorite in genetic research. J Zhejiang Univ-Sci B 19(10):739–749

Zuo LJ, Wang ZR, Tan YL et al (2016) piRNAs and their functions in the brain. Int J Hum Genet 16(1–2):53–60. https://doi.org/10.1080/09723757.2016.11886278

Recombinant DNA Technology

9

The concept of recombinant DNA (rDNA) technology was developed after the discovery of restriction enzymes by Werner Arber in bacteria that degrade foreign viral DNA molecules. This lead the geneticists to "cut" and "paste" DNA molecules, and novel restriction enzymes for cutting and pasting were discovered. The following are involved in the technology:

- Enzymes.
- Vectors.
- Host organism.

9.1 Enzymes

9.1.1 Restriction Enzymes

Restrictions enzymes are found in bacteria which were discovered by Werner Arber in 1968. Bacteria which are infected with viruses protect themselves by cutting randomly and invading foreign DNA molecules into an enormous numbers of fragments with enzymes called restriction enzymes. They are hydrolases that cleave phosphodiester bonds of double-stranded DNA at specific palindromic sites within the chain to produce 5′ PO4 and 3′ OH ends. Since these enzymes restrict the virus growth by binding to the viral DNA and cleaving it at recognition sequence, it is called restriction endonucleases. These recognition sequence are sites with high specificity through which bacteria defend its own DNA against attack by the restriction enzymes by methylating the corresponding sites of its own DNA.

There are three types of restriction enzymes. Type 1 and 3 recognize specific nonpalindromic sequences in the DNA, but cleave the chain at different sites away from the recognition site, thus producing DNA fragments of different length and ends. Example for Type 1 restriction endonuclease includes EcoK, EcoB and for

C. J. Betsy, C. Siva, *Fisheries Biotechnology and Bioinformatics*, https://doi.org/10.1007/978-981-99-6991-3_9

Type 3 includes Hind III. Type 2 recognize specific palindromic sequences that range generally from 4 to 8 nucleotides and cut the chain within the site, thus producing specific DNA fragments with known ends. Palindromes in DNA are sequences that read the same sequence of bases from either end (e.g., MALAYALAM). The restriction enzyme ECO R1 recognizes the GAATTC sequences and cuts the two strands at the sites shown producing staggered or sticky or cohesive 5′ and 3′ ends. Example for Type 2 includes Hindl, EcoRI, PvuII, Alul, Haelll, etc.

9.1.2 DNA Polymerase

DNA polymerases synthesize complementary nucleotide sequence on a template nucleotide strand. DNA polymerase I, isolated from *E. coli,* synthesizes a complementary strand on a template DNA in 5′ to 3′ direction. It has low level of exonuclease activity in both directions of DNA. It is used in probe preparation during DNA labeling.

9.1.3 Taq DNA Polymerase

Bacterium *Thermus aquaticus* from hot springs (94 °C) is used to isolate Taq DNA polymerase. It is used during polymerase chain reaction (PCR) for DNA amplification because of its high thermostability. It cannot proof read since it does not have 3′ to 5′ exonuclease activity.

9.1.4 Reverse Transcriptase

It is used to generate complementary DNA (cDNA) from an RNA template and to insert it into the chromosome of the host cell. It can be used to produce required genes from a complementary single-stranded DNA molecule synthesized on the mRNA template. Then DNA polymerase converts the single-stranded DNA into a double-stranded cDNA molecule.

9.1.5 DNA Ligase

Two DNA fragments from the same or different sources can be joined to produce recombinant DNA using DNA ligase. This is done by catalyzing the formation of phosphodiester bond between adjacent 3′ hydroxyl and 5′ phosphate termini of the two different DNA fragments. DNA ligase produced by T_4 bacteriophage is used which needs ATP and Mg^{++} for its activity. DNA ligase can join both single-stranded and double-stranded DNA fragments, DNA fragments having compatible staggered ends, and those with blunt ends. The rate of ligation of compatible staggered ends is 50-fold higher than that of blunt ends. It repairs nicks in double-stranded DNA molecules.

9.2 Vectors

Vectors are the carrier DNAs into which foreign DNAs or genes of interest are spliced to make a recombinant DNA. Vectors along with this foreign DNAs (i.e., recombinant DNA) are then introduced into appropriate host cell and are maintained for study or expression. These are also called as cloning vectors. They are used for obtaining millions of copies of cloned DNA segment. The cloned genes in these vectors are not expected to express themselves at transcription or translation level. Cloning vectors are used for creating genomic library or preparing the probes or genetic engineering experiments or other basic studies. Most cloning vectors were originally derived from naturally occurring extrachromosomal elements such as bacteriophages and plasmids.

9.2.1 Plasmids

They are found in *E. coil* during early 1950s by Joshua Lederberg from USA and William Hayes in England. Plasmids are genetic elements that are stably inherited without being a part of the chromosome of their host cell. They may be composed of DNA or RNA and may be linear or circular. It is a small DNA molecule within a cell that is physically separated from chromosomal DNA and can replicate independently (Fig. 9.1). Copy number is the number of molecules of a plasmid found in a single

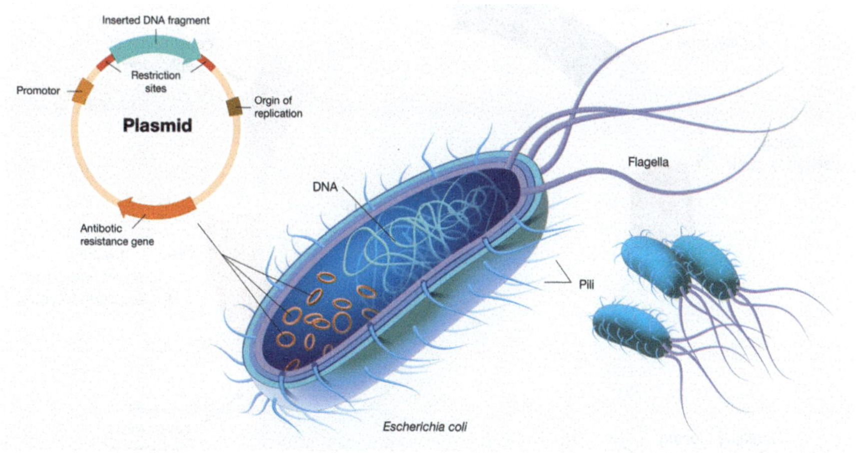

Fig. 9.1 Plasmids in *E. coli* (reproduced with permission from Darryl Leja, National Human Genome Research Institute)

bacterial cell. This ranges from 1 to more than 50 which is specific for a given plasmid residing in bacterial cell.

9.2.2 Types of Cloning Vectors

9.2.2.1 Shuttle Vectors

These vectors can be replicated in more than one organism and hence one particular gene can be expressed in different hosts. However, in order to be used in different host, these vectors must have separate origins of replication and selection mechanisms for each host. It also must have multiple cloning site for inserting the gene of interest (Fig. 9.2).

9.2.2.2 Bacteriophages as Cloning Vectors

Bacteriophage lambda which is a cloning vector is basically a bacterial virus that infects *E. coli*. Phage has a linear DNA (Fig. 9.3) molecule with complementary 12 bp long overhangs at each end known as cos sequences. When inserted in *E. coli*, these pair up and the cohesive ends are ligated together leading to circular version of the lambda genome. Only DNA molecules of between 37 kb and 52 kb can be stably packaged into the head of the lambda particle.

Small fragments of extra DNA may be inserted into the lambda genome by removing some of the lambda genome. The lambda can be divided into left, middle, and right portions. The genes for the structural proteins are present in the left portion, whereas the genes for replication and lysis are contained in the right portion. The non-essential genes are present in the middle portion which can be replaced with approximately 23 kb of foreign DNA (Fig. 9.4). However, the middle portion holds

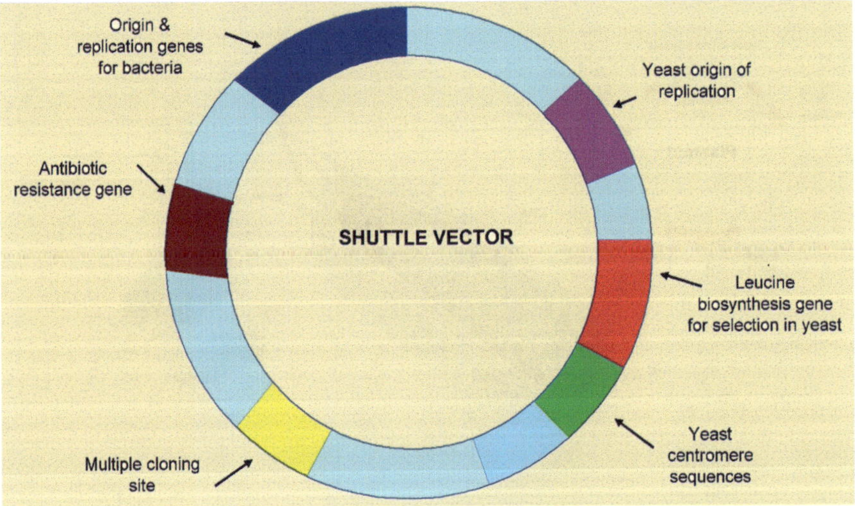

Fig. 9.2 Shuttle vector for yeast

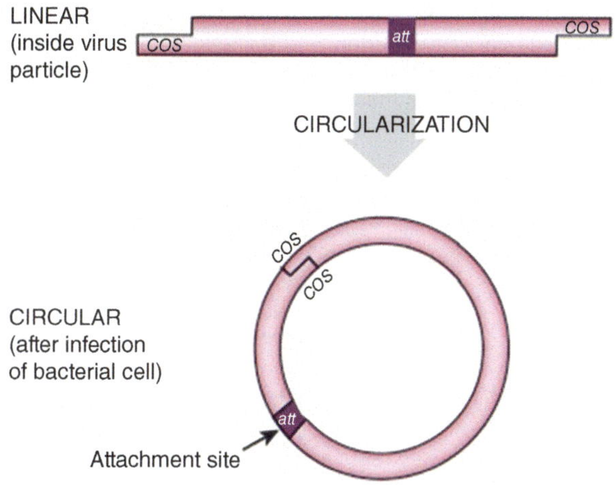

Fig. 9.3 Linear and circular genomes of Lambda (reproduced from Clark et al. 2019a with permission from Elsevier)

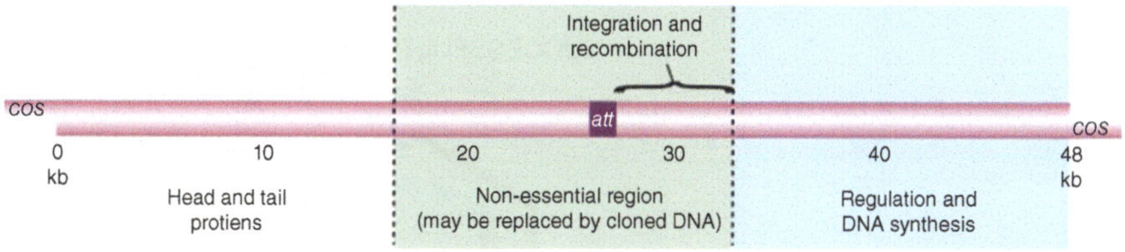

Fig. 9.4 Lambda replacement vector (reproduced from Clark et al. 2019b with permission from Elsevier)

the genes responsible for integration and recombination and hence, such lambda replacement vectors cannot integrate into the host chromosome and form lysogens by themselves. Therefore, a helper phage can be used to carry out the functions of integration and recombination.

If foreign DNA is inserted into the middle of lambda, the result is a linear DNA molecule with two cohesive ends. To get such constructs into an *E. coli,* in vitro packaging must be done (Fig. 9.5) where a mixture of lambda proteins is mixed with the recombinant lambda DNA to form phage particles. Infecting two separate *E. coli* cultures with two different defective lambda mutants generates the necessary lambda proteins. Each of the two mutants lacks an essential head protein and cannot form particles containing its own DNA. A mixture of the two lysates gives a full set of lambda proteins and when mixed with lambda DNA can generate infectious phage particles (Clark et al. 2019b).

9.2.2.3 Cosmids

Cosmids are the novel cloning vectors which possess properties of both plasmid and phage. They were first developed in 1978 by Barbara Hohn and John Collins. Cosmids contain cos site of phage (nucleotide sequence recognized for packing a phage DNA molecule into its protein capsule) plus essential feature of plasmid and

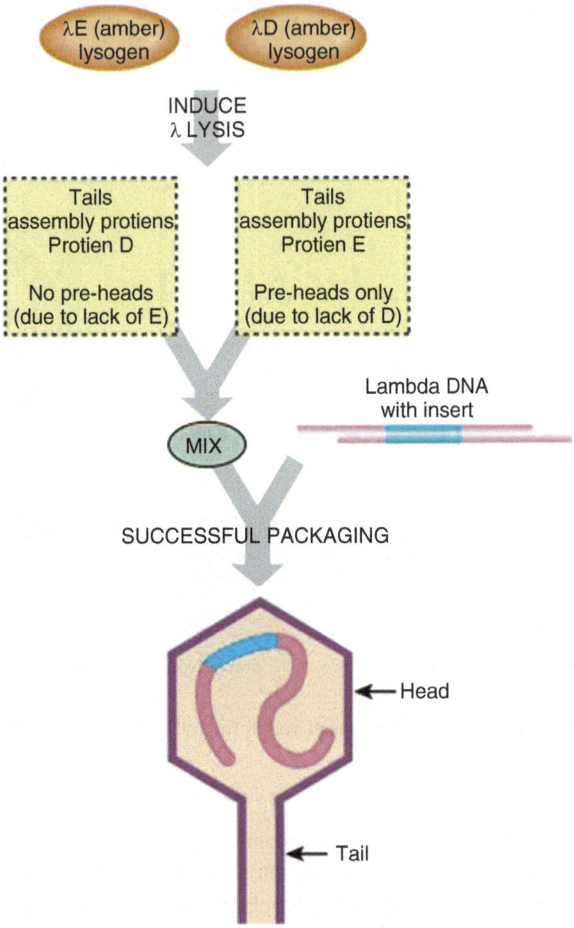

Fig. 9.5 In vitro packaging of Lambda replacement vector (reproduced from Clark et al. 2019b with permission from Elsevier)

several unique restriction sites for insertion of DNA to be cloned (Fig. 9.6). Cosmids can be perpetuated in bacteria in plasmid form, but can be purified by packing in vitro into phages. Advantage of using cosmid vector is that larger DNA can be cloned than what is possible with phage of plasmid.

In order to clone large pieces of DNA into cosmid vectors, the cosmid vector is linearized with cos site at each end. It is then cut with BamHI to generate sticky ends with the overhang sequence GATC. The genomic DNA is digested and partially digested with MboI, which also generates a GATC overhang. Partial digestion leaves some sites uncut and allows large segments of a genome to be isolated. These segments are mixed with the two halves of the cosmid and joined using ligase. The final constructs are packaged into lambda particles in vitro and are used to infect *E. coli* (Fig. 9.7) (Clark et al. 2019a).

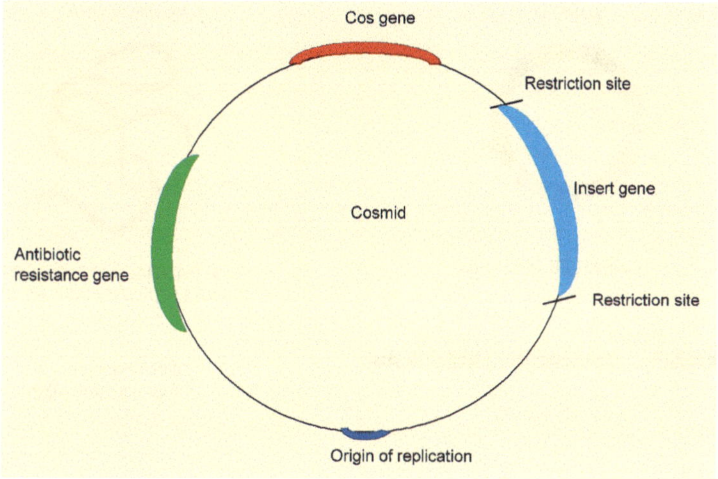

Fig. 9.6 Cosmid vectors

9.2.2.4 Yeast Cloning Vectors

Yeast cloning vectors are the carrier DNA molecules into yeasts. Yeasts are eukaryotes. Yeast artificial chromosomes (YACs) are most sophisticated yeast vectors. They have centromeric and telomeric region of a chromosome. These regions are needed to allow chromosome to be replicated in yeast cells. Due to origin of replication that is present, replication of DNA occurs. These elements are placed in single DNA fragment which can be used as vector to clone foreign DNA into yeasts (Fig. 9.8). The advantages of YAC include very large piece of DNA can be cloned. Only single copy of YAC is present per cell.

The YAC has two forms, a circular form for growing in bacteria, and a linear form for growing in yeast (Fig. 9.9). The circular form can be manipulated and grown in bacteria since it has a bacterial origin of replication and an antibiotic resistance gene. In order to use this in yeast, the circular form is isolated and linearized so that the yeast telomere sequences are on each end. This form can accommodate up to 2000 kb of cloned DNA inserted into its multiple cloning site (MCS) (Clark et al. 2019a).

9.2.2.5 Expression Vectors

Once a gene has been cloned into a vector, it may or may not be expressed. If both the structural gene and promoter were cloned on the same segment of DNA, the gene may well be expressed. If only the structural gene was cloned, then expression will depend on whether a promoter is provided by the plasmid. An expression vector contains sequences upstream of the cloned gene that control transcription and translation of the cloned gene. The expression vector uses the lac promoter, which is very strong, but inducible. To stimulate transcription, an artificial inducer molecule called IPTG is added. IPTG binds to the LacI repressor protein, which then

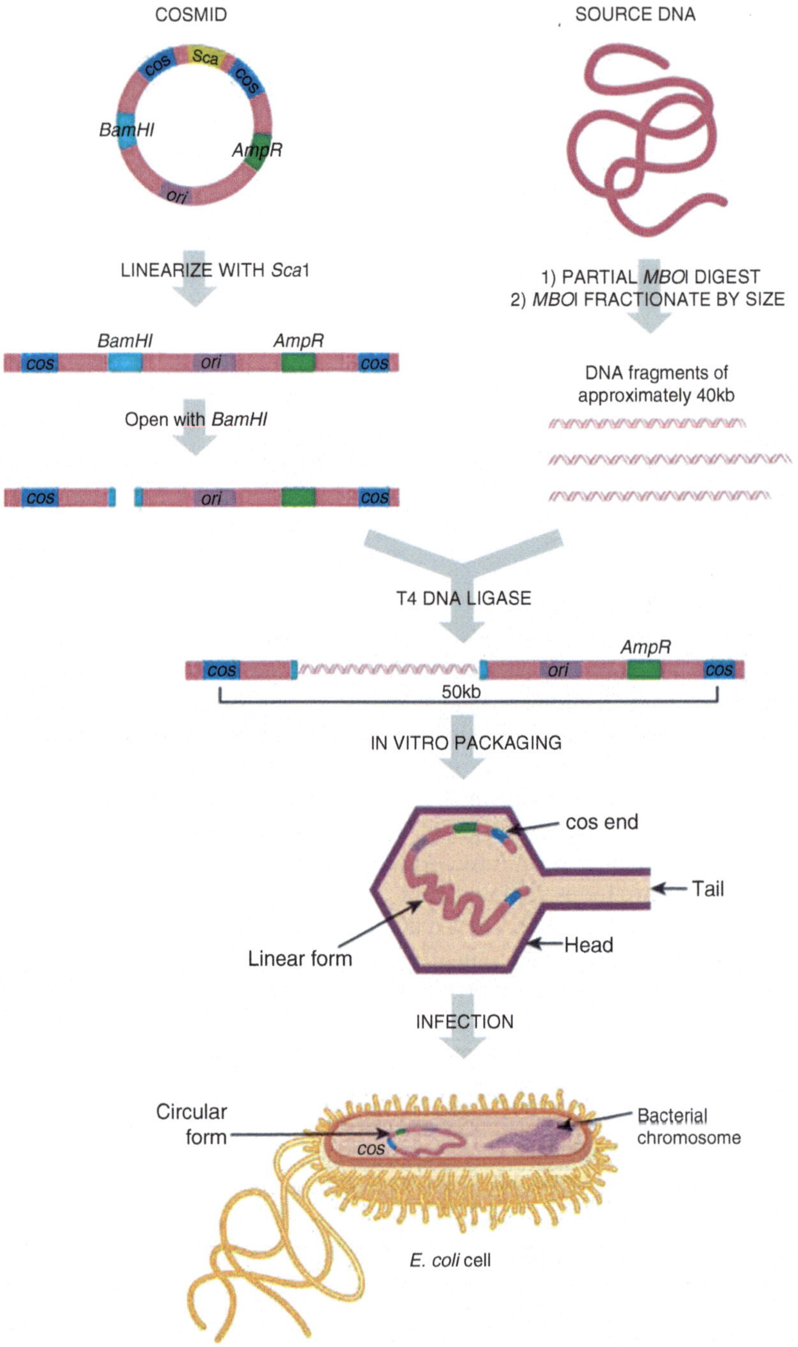

Fig. 9.7 Process of inserting cosmid vectors containing source DNA into *E. coli* (reproduced from Clark et al. 2019b with permission from Elsevier)

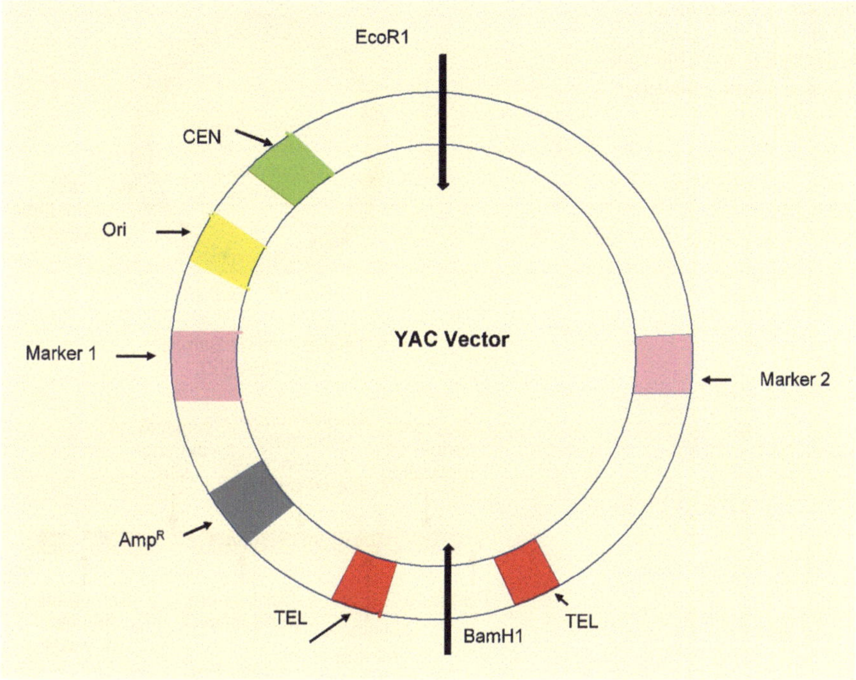

Fig. 9.8 YAC Vector

detaches from the DNA. This allows RNA polymerase to bind. Before IPTG is added to the culture, the LacI repressor prevents the cloned gene from being expressed (Fig. 9.10) (Clark et al. 2019a).

9.3 Host Organism

The host organism is the cell in which the recombinant DNA is introduced. It includes bacteria, fungi, and animal cells. To introduce vectors into hosts, techniques involving microinjection, biolistic, gene gun, alternate cooling, and heating, and calcium phosphate ions have been used.

9.4 Steps Involved

- Isolation of the genetic material (DNA).
- Cutting of DNA at specific locations.
- Isolation of desired DNA fragment.
- Amplification of gene of interest using PCR.
- Ligation of DNA fragment into a vector.

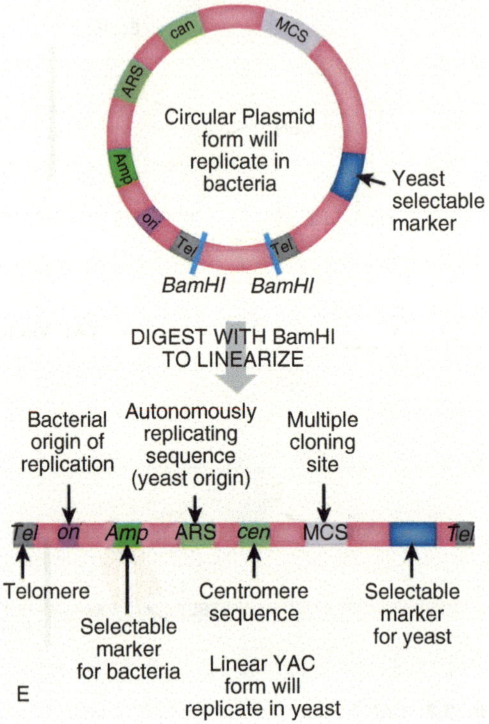

Fig. 9.9 Yeast artificial chromosome (reproduced from Clark et al. 2019b with permission from Elsevier)

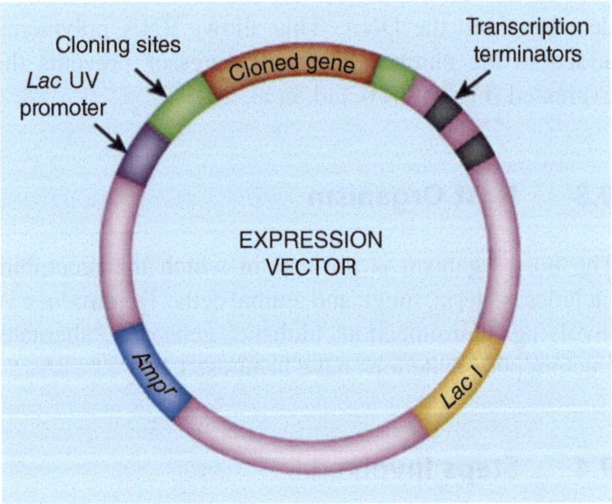

Fig. 9.10 Expression vectors (reproduced from Clark et al. 2019b with permission from Elsevier)

- Insertion of recombinant DNA into the host cell.
- Obtaining or culturing the foreign gene product.

9.4.1 Isolation of the Genetic Material

DNA is enclosed within the membranes. In order to release the DNA along with other macro-molecules (proteins, polysaccharides, and lipids), bacterial cells/plant or animal tissues are treated with the enzyme lysozyme (bacteria), cellulose (plant cells), chitinase (fungus), respectively. RNA can be extracted using ribonuclease and for protein, protease is used. Purified DNA will precipitate out, after the addition of chilled ethanol. This can be seen as collection of fine threads in the suspension.

9.4.2 Cutting of DNA at Specific Locations

The target DNA is cut with the help of specific restriction enzymes.

9.4.3 Isolation of Desired DNA Fragment

Using agarose gel electrophoresis, the activity of the restriction enzymes can be checked. Since, the DNA is negatively charged, it moves towards the positive electrode or anode and DNA tends to separates out in this process. After that the desired DNA fragment is eluted out.

9.4.4 Amplification of Gene of Interest Using PCR

The gene of interest is amplified using PCR. Polymerase Chain Reaction (PCR) is best defined as the DNA replication in vitro.

9.4.5 Ligation of DNA Fragment into a Vector

This process requires a vector DNA and a source DNA. In order to obtain sticky ends, both of these should cut with the same endonuclease. After which both are ligated by mixing vector DNA, gene of interest and enzyme DNA ligase form the recombinant DNA/hybrid DNA.

9.4.6 Insertion of Recombinant DNA into the Host Cell

This can occur by several methods, before which the recipient cells are made competent to receive the DNA. If a recombinant DNA bearing gene for resistance to an antibiotic (e.g., ampicillin) is transferred into *E. coli* cells, the host cells become transformed into ampicillin resistant cells.

9.4.7 Culturing the Foreign Gene Product

When a piece of foreign DNA is inserted into a cloning vector and transferred into a bacterial cell, the foreign DNA gets multiplied and produces desirable protein expression.

9.4.8 Cloning Foreign DNA into Plasmid DNA

Cloning is generally done in yeast cells since in *E. coli* splicing of mRNA cannot be done. Restriction endonuclease digests the cell's total genome and produces sticky ends to get the foreign DNA. The same enzyme is used to cut open the circular plasmid. The cut fragments of the whole genome and plasmids are annealed and ligated to get recombinant plasmids in which at least one or two carry the desired gene. The highly heterogenous mixture of recombinant and parental plasmids so produced is introduced into the yeast cells by the process called transformation. The permeability of yeast cell wall is increased by treating it with high concentration of Ca^{++}. The cells are then placed on a selective medium containing an antibiotic to which the plasmid carries the resistance gene. The transformed cells grow and produce cell colonies on the plate, each colony being formed by a single transformed cell (Fig. 9.11).

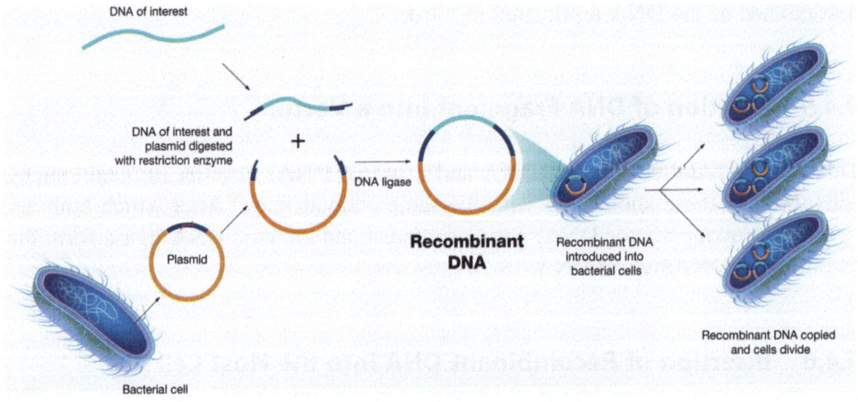

Fig. 9.11 Steps involved in recombinant DNA technology (reproduced with permission from Darryl Leja, National Human Genome Research Institute)

References

Clark DP, Pazdernik NJ, McGehee MR (2019a) Cloning genes for synthetic biology. In: Molecular biology, pp 199–239. https://doi.org/10.1016/b978-0-12-813288-3.00

Clark DP, Pazdernik NJ, McGehee MR (2019b) Regulation of transcription in prokaryotes. In: Molecular biology, pp 522–559. https://doi.org/10.1016/b978-0-12-813288-3.000

10 Transgenic Fish Production

An organism that has a foreign or modified gene in its genome is called transgenic or GMO. A genetically modified organism (GMO) is an organism whose genetic material has been altered using genetic engineering techniques. Organisms that have been genetically modified include microorganisms such as bacteria and yeast, insects, plants, fish, and mammals. The first recorded instances of production of transgenic in aquatic species are those of rainbow trout and goldfish. Since then over 35 species have been genetically engineered. The following are the steps involved:

1. Selection of fish species
2. Preparation of a gene construct
3. Isolation of specific gene
4. Cloning the gene of interest
5. Gene transfer
6. Expression of transgenes
7. Inheritance of transgenes through further generation

10.1 Selection of Fish Species

Several different fish species including channel catfish, common carp, goldfish, Japanese medaka, loach, northern pike, rainbow trout, salmon, tilapia, walleye, and zebrafish have been used as experimental animals for production of transgenic fish. Depending on the purpose of the transgenic fish studies, the embryos of some fish species are better suited for gene transfer studies than the others. For example, Japanese medaka (*Oryzias latipes*) and zebrafish (*Brachydanio rerio*) have short life cycles (3 months from hatching to mature adults) and produce hundreds of eggs on a regular basis without exhibiting a seasonal breeding cycle and can be maintained easily in the laboratory for 2–3 years. Eggs from these two fish species are relatively large (diameter: 0.7–1.5 mm) and possess very thin, semi-transparent chorions, a

C. J. Betsy, C. Siva, *Fisheries Biotechnology and Bioinformatics*,
https://doi.org/10.1007/978-981-99-6991-3_10

feature that permits easy microinjection of DNA into fertilized eggs. Furthermore, inbred lines and various morphological mutants of both fish species are available. Therefore, these fish species are suitable candidates for conducting gene transfer experiments for (a) studying developmental regulation of gene expression and gene action; (b) identifying regulatory elements that regulate the expression of a gene; (c) measuring the activities of promoters; and (d) producing transgenic models for environmental toxicology. However, a major drawback of these two fish species is their small body size that makes them unsuitable for some endocrinological or biochemical analyses.

Channel catfish, common carp, rainbow trout, and salmon are commonly used as large body size model fish species in transgenic fish studies. Since the endocrinology, reproductive biology, and basic physiology of these fish species have been well worked out, they are well suited for conducting studies on comparative endocrinology and aquaculture applications. However, the long maturation time of these fish species and a single spawning cycle per year will limit research progress in this field.

Loach, killifish, goldfish, and tilapia are the third group of model fish species suitable for conducting gene transfer studies since their body sizes are large enough for most biochemical and endocrinological studies. Furthermore, shorter maturation times, as compared to catfish, rainbow trout, or salmon allow easier manipulation of transgenic progeny. Unfortunately, the lack of a well-defined genetic background and asynchronous reproductive behavior of these fish species render them less amenable to gene transfer studies.

10.2 Preparation of Transgene Constructs

Transgenes used in producing transgenic fish for basic research or application are recombinant gene constructs that produce gene products at appropriate levels in the desired tissue(s) at the desired time(s). Therefore, the prototype of a transgene is usually constructed in a plasmid to contain an appropriate promoter/enhancer element and a structural gene sequence. Depending on the purpose of gene transfer studies, transgenes are grouped into three main types: (1) gain-of-function, (2) reporter function, and (3) loss-of-function.

The gain-of-function transgenes are designed to add new functions to the transgenic individuals or to facilitate the identification of the transgenic individuals if the genes are expressed properly in the transgenic individuals. Transgenes containing the structural genes of mammalian and fish growth hormones (GH) genes or their cDNAs fused to functional promoters such as chicken and fish β-actin gene promoters are examples of the gain-of-function transgene constructs. Expression of the GH transgenes in transgenic individuals will result in increased production of GH and ultimate growth enhancement.

Bacterial chloramphenicol acetyl transferase (CAT), β-galactosidase, or luciferase genes fused to functional promoters are examples of transgenes with reporter function. These reporter function transgenes are commonly used to identify the success of gene transfer effort. A more important function of a reporter gene is used to identify and measure the strength of a promoter/enhancer element. In this case, the structural gene of the CAT, β-galactosidase, or luciferase gene is fused to a promoter/enhancer element in question. Following gene transfer, the expression of the reporter gene activity is used to determine the transcriptional regulatory sequence of a gene or the strength of a promoter.

The "loss-of-function" transgenes are constructed for interfering with the expression of host genes. These genes might encode an antisense RNA to interfere with the post-transcriptional process or translation of endogenous mRNAs. Alternatively, these genes might encode a catalytic RNA (a ribozyme) that can cleave specific mRNAs and thereby cancel the production of the normal gene product. Although these genes have not yet been introduced into a fish model, they could potentially be employed to produce disease resistant transgenic brood-stocks for aquaculture or transgenic model fish defective in a particular gene product for basic research.

10.3 Isolation of the Gene of Interest

Usually the gene of interest will already be available as an element of a "library" of short sections of the total genome of the donor strain or species. If this is the case, the procedure followed is to multiply the gene using the PCR reaction. If, however, the gene is to be taken from a genome not previously investigated, a more complex procedure will need to be followed. PCR enables the gene in both the cases noted above to be multiplied to the level of several million copies needed for the generation of the construct.

10.3.1 Genes of Interest

10.3.1.1 Growth Hormone (GH) Gene

GH is produced by the anterior lobe of the pituitary. GH increases growth by stimulating appetite and improving the food conversion efficiency. GH is a protein hormone having a molecular weight of about 22 kilodaltons. Transgenic fish carrying GH gene will produce growth hormone endogenously by passing the necessity of exogenous hormone treatment. GH gene has been cloned in some fishes either from the genomic library or from the cDNA library.

10.3.1.2 Anti-freeze Protein Genes

Production of cold resistant fish variety is useful for establishing aquaculture industry in the temperate region, where water gets frozen during winter. The gene responsible for imparting cold resistance was cloned from winter flounder (*Pseudopleuronectes americanus*), which lives in the polar sea. This species avoids freezing of its blood even at −7 °C temperatures by producing a set of an anti-freeze proteins (AFP). AFP are produced in the liver and exported to the blood stream.

10.3.1.3 Disease Resistance

Fish has poorly developed antibody dependent immunity. Efforts to produce disease resistance in fish stocks by transgenesis have begun recently. The potential of rainbow trout lysozyme gene as a bacterial inhibitor was assessed in Atlantic salmon.

10.4 Cloning the Gene of Interest

When many copies of the target gene have been generated, the gene is placed in a "construct." Once the gene of interest has been ligated enzymatically into the construct, this whole complex is ligated into bacterial plasmids, which act as "production vectors" and enable the gene to be replicated many times within the bacterial cells. The bacteria are then plated out, and the reporter genes will help to know whether the vector has been taken up by the bacterial cells. This usually involves some color change in the colonies containing inserted DNA. The many times amplified DNA construct is then enzymatically cut out of the plasmids (after these have been removed from the bacterial cells), and it is ready to be used for insertion into eggs of the host species.

A construct is a piece of DNA which functions as the vehicle or vector carrying the target gene into the recipient organism. It has several different regions like

- Promoter region—controls the activity of the target gene
- A region where the target DNA is inserted
- Reporter gene to enable one to ascertain whether the target has combined successfully with the construct and a termination sequence

10.5 Gene Transfer

Techniques such as calcium phosphate precipitation, direct microinjection, lipofection, retrovirus infection, electroporation, and biolistic bombardment have been used to introduce foreign DNA into animal cells, plant cells, and germ lines of mammals and other vertebrates. Among these methods, direct microinjection and electroporation of DNA into newly fertilized eggs have been proven to be the most reliable methods of gene transfer in fish systems.

10.5.1 Microinjection

Microinjection of foreign DNA into newly fertilized eggs was first developed for the production of transgenic mice in the early 1980s. Since 1985, this technique has been adopted for introducing transgenes into Atlantic salmon, common carp, catfish, goldfish, loach, medaka. Eggs and sperm are collected in separate, dry containers. Fertilization is initiated by adding water and sperm to the eggs, with gentle stirring to enhance fertilization. Fertilized eggs are then microinjected within the first few hours after fertilization. The injection apparatus consists of a dissecting stereo microscope and two micromanipulators, one with a micro-glass-needle for delivering transgenes and the other with a micropipette for holding fish embryos in place. About 10^6–10^8 molecules of a linearized transgene (with or without plasmid DNA) in about 20 nL are injected into the egg cytoplasm. Following injection, the embryos are incubated in water until hatching. Depending on the fish species, the survival rate of injected fish embryos ranges from 35% to 80% while the rate of DNA integration ranges from 10% to 70% in the survivors. The tough chorions of the fertilized eggs in some fish species, e.g., rainbow trout and Atlantic salmon, can frequently make insertion of glass needles in microinjection difficult which can be overcome by any one of the following methods:

(a) Inserting the injection needles through the micropyle
(b) Making an opening on the egg chorions by microsurgery
(c) Removing the chorion by mechanical or enzymatic means
(d) Reducing chorion hardening by initiating fertilization in a solution containing 1 mM glutathione
(e) Injecting the unfertilized eggs directly

10.5.2 Electroporation

Electroporation utilizes a series of short electrical pulses to permeate cell membranes, thereby permitting the entry of DNA molecules into embryos. The patterns of electrical pulses can be emitted in a single pulse of exponential decay form (i.e., exponential decay generator) or high frequencies multiple peaks of square waves (i.e., square wave generator). Rate of DNA integration in electroporated embryos is on the order of 20% or higher in the survivors. Both methods are not successful in marine fish and invertebrates.

10.5.3 Transgenic Mosaicism

Embryonic development occurs at very fast rate, and exogenous DNA in fishes takes hours to get integrate into the host genome. By this time, several round of replication should have already been over. As a result some cell types contain transgene while others do not, this is called transgenic mosaicism.

10.6 Identification of Transgenic Fish

- Dot blot hybridization
- Southern blot hybridization
- PCR

10.7 Expression of Transgenes

- RNA northern or dot blot hybridization
- RNase protection assay
- Reverse transcription polymerase chain reaction (RT/PCR)
- Immunoblotting assay
- Other biochemical assays

10.8 Inheritance of Transgenes

A fish which expresses the target gene at an acceptable level may not be able to transmit the gene to progeny. This is because many transgenics are mosaic individuals and unless the gonads are included in the tissues possessing the transgene, the transgenic animals will not breed true. Appropriate breeding tests must, therefore, be carried out. To determine whether the transgene is transmitted to the subsequent generation, P1 transgenic individuals are mated to non-transgenic individuals and the progeny is assayed for the presence of transgenes by the PCR assay method.

10.9 Advantages

- Increased growth rate
- Increased market size
- Decreased dress out percentage
- Improved feed conversion efficiency
- Improved cold tolerance/freeze resistance
- Increased brood stock fecundity
- Controlled smolting, reproduction, and sex
- Reduced aggression
- Improved disease resistance
- Production of new color

10.10 Risks

- Escape into wild
- Ecological disruption—genetically modified fish survives better when there is a shortage of food, and benefit more than non-transgenic fish from increasing water temperature
- Where food is abundant, transgenic fish can quickly grow and have the potential to displace the natural one
- Genetic diversity of natural population will be disturbed

Animal Cell Culture

11

Cell culture refers to the process by which cells are grown in a controlled artificial environment. In a cell culture technique, cells are removed from an animal or a plant and grown subsequently in a favorable environment. For animal cell culture, the cells are taken from the organ of an experimental animal. The cells may be removed directly or by mechanical or enzymatic action. The cells can also be obtained by previously made cell line or cell strain. Examples of cells used to culture are fibroblast, lymphocytes, cells from cardiac and skeletal tissues, cells from liver, breast, skin, and kidney and different types of tumor cells.

11.1 Types of Animal Cell Culture

11.1.1 Based on the Number of Cell Division

11.1.1.1 Primary Cell Culture

This is the cell culture obtained straight from the cells of a host tissue. The cells dissociated from the parental tissue are grown on a suitable container and the culture thus obtained is called primary cell culture. Such culture comprises mostly heterogeneous cells and most of the cells divide only for a limited time. However, these cells are much similar to their parents. Depending on their origin, primary cells grow either as an adherent monolayer or in a suspension.

Adherent Cells

These cells are anchorage dependent and propagate as a monolayer. These cells need to be attached to a solid or semi-solid substrate for proliferation. These adhere to the culture vessel with the use of an extracellular matrix which is generally derived from tissues of organs that are immobile and embedded in a network of connective tissue. Fibroblasts and epithelial cells are of such types.

C. J. Betsy, C. Siva, *Fisheries Biotechnology and Bioinformatics*,
https://doi.org/10.1007/978-981-99-6991-3_11

When the bottom of the culture vessel is covered with a continuous layer of cells, usually one cell in thickness, these are known as monolayer cultures. The majority of continuous cell lines grow as monolayers. As being single layers, such cells can be transferred directly to a cover slip to examine under microscope.

Suspension Cells

Suspension cells do not attach to the surface of the culture vessels. These cells are also called anchorage independent or non-adherent cells which can be grown floating in the culture medium. Hematopoietic stem cells (derived from blood, spleen and bone marrow) and tumor cells can be grown in suspension. These cells grow much faster which do not require the frequent replacement of the medium and can be easily maintained. These are of homogeneous types, and enzyme treatment is not required for the dissociation of cells; similarly these cultures have short lag period.

Confluent Culture and the Necessity of Subculture

After the cells are isolated from the tissue and proliferated under the appropriate conditions, they occupy all of the available substrate, i.e., reach confluence. For a few days, it can become too crowded for their container and this can be detrimental to their growth, generally leading to cell death if left for a long time. The cells thus have to be subcultured, i.e., a portion of cells is transferred to a new vessel with fresh growth medium which provides more space and nutrients for the continual growth of both portions of cells, hence subculture keeps cells healthy in a growing state.

A passage number refers to how many times a cell line has been subcultured. In contrast with the population doubling level where the specific number of cells involved is not relevant, it simply gives a general indication of how old the cells may be for various assays.

11.1.1.2 Secondary Cell Culture and Cell Line

When a primary culture is subcultured, it is known as secondary culture or cell line or sub-clone. The process involves removing the growth media and disassociating the adhered cells (usually enzymatically).

Sub-culturing of primary cells to different divisions leads to the generation of cell lines. During the passage, cells with the highest growth capacity predominate, resulting in a degree of genotypic and phenotypic uniformity in the population. However, as they are subcultured serially, they become different from the original cell.

11.1.2 Based on the Life Span of Culture

11.1.2.1 Finite Cell Lines

The cell lines which go through a limited number of cell division having a limited life span are known as finite cell lines. The cells passage several times and then lose their ability to proliferate, which is a genetically determined event known as senescence. Cell lines derived from primary cultures of normal cells are finite cell lines.

11.1.2.2 Continuous Cell Lines

When a finite cell line undergoes transformation and acquires the ability to divide indefinitely, it becomes a continuous cell line. Such transformation/mutation can occur spontaneously or can be chemically or virally induced or from the establishment of cell cultures from malignant tissue. Cell cultures prepared in this way can be subcultured and grown indefinitely as permanent cell lines and are immortal.

These cells are less adherent, fast growing, less fastidious in their nutritional requirements, able to grow up to higher cell density and different in phenotypes from the original tissue. Such cells grow more in suspension. They also have a tendency to grow on top of each other in multilayers on culture-vessel surfaces.

11.2 Cell Strain

Lineage of cells originated from the primary culture is called strain. These are either derived from a primary culture or a cell line by the positive selection or cloning of cells having specific properties or characteristics. A cell strain often acquires additional genetic changes subsequent to the initiation of the parent line.

11.2.1 Methods

11.2.1.1 Growth Requirements

The culture media used for cell cultures are generally quite complex, and culture condition widely varies for each cell type. However, media generally include amino acids, vitamins, salts (maintain osmotic pressure), glucose, a bicarbonate buffer system (maintains a pH between 7.2 and 7.4), growth factors, hormones, O_2, and CO_2. To obtain best growth, addition of a small amount of blood serum is usually necessary, and several antibiotics like penicillin and streptomycin are added to prevent bacterial contamination.

Temperature varies on the type of host cell. Most mammalian cells are maintained at 37 °C for optimal growth, while cells derived from cold-blooded animals tolerate a wider temperature range (i.e., 15–26 °C). Actively growing cells of log phase should be used which divide rapidly during culture.

11.2.1.2 Process to Obtain Primary Cell Culture

Primary cell cultures are prepared from fresh tissues. Pieces of tissues from the organ are removed aseptically; which are usually minced with a sharp sterile razor and dissociated by proteolytic enzymes such as trypsin that break apart the intercellular cement. The obtained cell suspension is then washed with a physiological buffer to remove the proteolytic enzymes used. The cell suspension is spread out on the bottom of a flat surface, such as a bottle or a Petri dish. This thin layer of cells adhering to the glass or plastic dish is overlaid with a suitable culture medium and is incubated at a suitable temperature.

11.2.1.3 Aseptic Techniques

Bacterial infections, like Mycoplasma and fungal infections, commonly occur in cell culture creating a problem to identify and eliminate. Thus, all cell culture work is done in a sterile environment with proper aseptic techniques. Work should be done in laminar flow with the constant unidirectional flow of HEPA filtered air over the work area. All the material, solutions, and the whole atmosphere should be of contamination-free.

11.2.1.4 Cryopreservation

If a surplus of cells is available from sub-culturing, they should be treated with the appropriate protective agent (e.g., DMSO or glycerol) and stored at temperatures below −130 °C until they are needed. This stores cell stocks and prevents original cell from being lost due to unexpected equipment failure or biological contaminations. It also prevents finite cells from reaching senescence and minimizes risks of changes in long-term cultures. When thawing the cells, the frozen tube of cells is warmed quickly in warm water, rinsed with medium and serum, and then added into culture containers once suspended in the appropriate media.

11.3 Applications of Cell Line

11.3.1 Vaccines Production

One of the most important uses of cell culture is in research and production of vaccines. The ability to grow large amounts of virus in cell culture eventually led to the creation of the polio vaccine, and cells are still used today on a large scale to produce vaccines for many other diseases, like rabies, chickenpox, hepatitis B, and measles. In early times, researchers had to use live animals to grow poliovirus, but due to the development of cell culture techniques, they were able to achieve much greater control over virus production and on a much larger scale which eventually develop vaccines and various treatments. However, continuous cell lines are not used in virus production for human vaccines as these are derived from malignant tissue or possess malignant characteristics.

11.3.2 Virus Cultivation and Study

Cell culture is widely used for the propagation of viruses as it is convenient, economic, and easy to handle compared to other animals. It is easy to observe cytopathic effects and easy to select particular cells on which the virus grow as well as to study the infectious cycle. Cell lines are convenient for virus research because cell material is continuously available. Continuous cell lines have been extremely useful in cultivating many viruses previously difficult or impossible to grow.

11.3.3 Cellular and Molecular Biology

Cell culture is one of the major tools used in cellular and molecular biology, providing excellent model systems for studying the normal physiology and biochemistry of cells (e.g., metabolic studies, aging), the effects of different toxic compounds on the cells, and mutagenesis and carcinogenesis. The major advantage of using cell culture for any of these applications is the consistency and reproducibility of results that can be obtained from using a batch of clonal cells.

11.3.4 Cancer Research

Normal cells can be transformed into cancer cells by methods including radiation, chemicals, and viruses. These cells can then be used to study cancer more closely and to test potential new treatments.

11.3.5 Gene Therapy

Cells having a functional gene can be replaced to cells which are having non-functional gene, and for which the cell culture technique is used.

11.3.6 Immunological Studies

Cell culture techniques are used to know the working of various immune cells, cytokines, lymphoid cells, and interaction between disease-causing agents and the host cells.

11.3.7 Others

Cell lines are also used in in vitro fertilization (IVF) technology, recombinant protein, and drug selection and improvement.

12 Hybridoma Technology

Monoclonal antibodies are produced by hybridoma technology. The term hybridoma is applied to fused cells resulting due to fusion of lymphocytes and tumor cell. Usually an antibody-producing B lymphocytes (e.g., spleen cell of mouse immunized with RBCs from sheep) and a single myeloma cell (e.g., bone marrow tumor cell) can be adopted to grow for infinite time in culture. The fused product derives the ability to produce large amount of pure antibodies as lymphocytes and ability to grow or multiply indefinitely like tumor cell.

In 1964, Littlefield developed a way to isolate hybrid cells from two parent cell lines using the hypoxanthine-aminopterin-thymidine (HAT) selection media. In 1975, hybridoma technology was first discovered by Georges Kohler and Cesar Milstein. They were also awarded Nobel Prize along with Niels Kaj Jerne in Physiology and Medicine field during 1984. However, Milstein produced the first monoclonal antibodies in 1990.

12.1 Monoclonal Antibodies (MAbs)

- Antibodies are glycoprotein synthesized in blood against specific antigens to combat and give immunity. Such antibodies are heterogenous and are polyclonal antibodies. Therefore they do not have characteristics of specificity.
- If a specific lymphocyte after isolation and culture in vitro becomes capable of producing a single type of antibody which bears specificity against specific antigen, it is known as monoclonal antibody. These monoclonal antibodies are derived from a single clone of cell which recognize only one kind of antigen. Monoclonal antibodies are produced against a variety of proteins, glycoproteins, glycolipids, nucleic acids, and chemically defined groups linked to protein carriers.

C. J. Betsy, C. Siva, *Fisheries Biotechnology and Bioinformatics*,
https://doi.org/10.1007/978-981-99-6991-3_12

12.2 Production of MAb

In order to produce MAb, B cells and myeloma cells are used. B cells are $HGPRT^+$ and are able to produce Ab but they are mortal cells. Myeloma cells have lost the ability to synthesize hypoxanthine-guanine phosphoribosyl transferase (HGPRT), an enzyme necessary for the salvage synthesis of nucleic acids. By exposing cells to aminopterin (a folic acid analogue, which inhibits dihydrofolate reductase, DHFR), they are unable to use the de novo pathway and become fully auxotrophic for nucleic acids requiring supplementation to survive. Hence myeloma cells are $HGPRT^-$, Ig^- and are immortal.

MAbs are typically made by fusing myeloma cells with the spleen cells from a mouse that has been immunized with the desired antigen. Polyethylene glycol is used to fuse adjacent plasma membranes. A selective medium in which only fused cells can grow is used. The selective culture medium is called HAT medium because it contains hypoxanthine, aminopterin, and thymidine. This medium is selective for fused (hybridoma) cells. Unfused myeloma cells cannot grow because they lack HGPRT and thus cannot replicate their DNA. Unfused spleen cells cannot grow indefinitely because of their limited life span. Only fused hybrid cells, referred to as hybridomas, are able to grow indefinitely in the media because the spleen cell partner supplies HGPRT and the myeloma partner has traits that make it immortal like a cancer cell. This mixture of cells is then diluted, and clones are grown from single parent cells on microtitre wells. The antibodies secreted by the different clones are then assayed for their ability to bind to the antigen (with a test such as ELISA or antigen microarray assay) or immuno-dot blot. The most productive and stable clone is then selected for future use.

The hybridomas can be grown indefinitely in a suitable cell culture medium. They can also be injected into mice (in the peritoneal cavity, surrounding the gut). There, they produce tumors secreting an antibody-rich fluid called ascites fluid (Fig. 12.1). The medium must be enriched during in vitro selection to further favor hybridoma growth. This can be achieved by the use of a layer of feeder fibrocyte cells or supplement medium such as briclone. Culture medium conditioned by macrophages can also be used. Production in cell culture is usually preferred as the ascites technique is painful to the animal and is considered unethical.

12.2.1 Steps in Production of MAb

12.2.1.1 Immunization of Mice

Mice are immunized every 2–3 weeks with an antigen that is prepared for injection either by emulsifying the antigen with Freund's adjuvant or other adjuvants or by homogenizing a gel slice that contains the antigen. Intact cells, whole membranes, and microorganisms are sometimes used as immunogens. When a sufficient antibody titre is reached in serum, immunized mice are euthanized and the spleen is removed to use as a source of cells for fusion with myeloma cells.

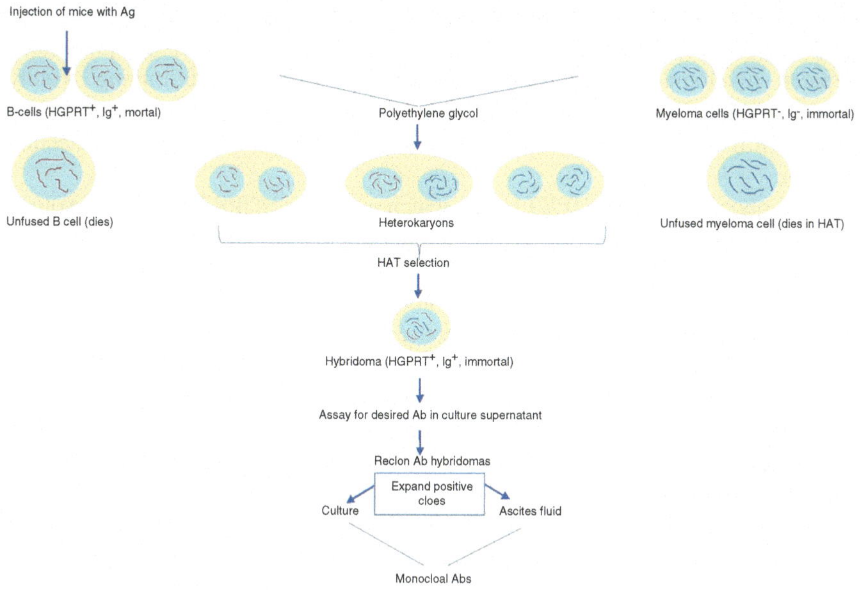

Fig. 12.1 Steps involved in the production of monoclonal antibodies

12.2.1.2 Screening of Mice for Antibody Production

After several weeks of immunization, blood samples are obtained from mice for measurement of serum antibodies. Serum antibody titre is determined with various techniques, such as enzyme-linked immunosorbent assay (ELISA) and flow cytometry. If the antibody titre is high, cell fusion can be performed. If the titre is too low, mice can be boosted until an adequate response is achieved, as determined by repeated blood sampling. When the antibody titre is high enough, mice are commonly boosted by injecting antigen without adjuvant intraperitoneally or intravenously (via the tail veins) 3 days before fusion but 2 weeks after the previous immunization. Then the mice are euthanized, and their spleens are removed for in vitro hybridoma cell production.

12.2.1.3 Preparation of Myeloma Cells

Fusing antibody-producing spleen cells, which have a limited life span, with cells derived from an immortal tumor of lymphocytes (myeloma) results in a hybridoma that is capable of unlimited growth. Myeloma cells are immortalized cells that are cultured with 8-azaguanine to ensure their sensitivity to the hypoxanthine-aminopterin thymidine (HAT) selection medium used after cell fusion. A week before cell fusion, myeloma cells are grown in 8-azaguanine. Cells must have high viability and rapid growth. The HAT medium allows only the fused cells to survive in culture.

12.2.1.4 Fusion of Myeloma Cells with Immune Spleen Cells

Single spleen cells from the immunized mouse are fused with the previously prepared myeloma cells. Fusion is accomplished by co-centrifuging freshly harvested spleen cells and myeloma cells in polyethylene glycol, a substance that causes cell membranes to fuse. Only fused cells will grow in the special selection medium.

B-lymphocytes contain HPRT1 gene which codes for HGPRT which is involved in synthesis of nucleotides from hypoxanthine present in culture medium. Therefore, B cells can grow in HAT medium. But myeloma cell lacks HPRT1 gene so it does not produce HGPRT enzyme and it does not grow in HAT medium. The myeloma cell fused with another myeloma cell or those do not fuse at all die in HAT medium since they do not utilize hypoxanthine. Similarly, B cell that fuse with another B cell or those do not fuse at all die eventually because they do not have the capacity to divide indefinitely. So only hybridoma cell can survive and divide in HAT medium. Screening is done to select hybridoma cells which are the desired cell for monoclonal antibodies production.

The cells are then distributed to 96-well plates containing feeder cells derived from saline peritoneal washes of mice. Feeder cells are believed to supply growth factors that promote growth of the hybridoma cells. Commercial preparations that result from the collection of media supporting the growth of cultured cells and contain growth factors are available that can be used in lieu of mouse-derived feeder cells. It is also possible to use murine bone marrow-derived macrophages as feeder cells.

12.2.1.5 Isolation and Cloning of Hybridoma Cell with Single Specificity

At this step, new, small clusters of hybridoma cells from the 96-well plates can be grown in tissue culture followed by selection for antigen binding or grown by the mouse ascites method with cloning at a later time. Cloning by limiting dilution at this time ensures that a majority of wells each contain at most a single clone. Considerable judgment is necessary at this stage to select hybridomas capable of expansion versus total loss of the cell fusion product due to under population or inadequate in vitro growth at high dilution. In some instances, the secreted antibodies are toxic to fragile cells maintained in vitro. Optimizing the mouse ascites expansion method at this stage can save the cells.

12.3 Application of Monoclonal Antibodies (MAbs) in Aquaculture

MAbs are being employed in disease diagnosis, pathogen classification, epidemiological analysis, and development of vaccines. Because of their specificity, MAbs are more pure than the polyclonal antibodies of conventional techniques and potentially more effective than contemporary drugs in fighting disease. Drugs often attack the body's own cells in addition to the foreign molecule, producing negative side effects such as nausea. MAbs will bind to only the specific target molecule, without any unwanted side effects. Once monoclonal antibodies for a given substance have been produced, they can be used to detect the presence of this substance. MAb-based immunodiagnostic kits are available for several pathogens. ELISA and immuno-dot kits based on MAb are being used by farmers. MAbs have also been extensively used in detection of epitopes involved in pathogenesis for development of subunit vaccines. MAbs to *A. hydrophila*, EUS fungus, and white spot virus of shrimp have been produced and being used in diagnosis.

13 Cryopreservation Technology in Fishes

Cryopreservation of cells is a process wherein biological materials like living cells or tissues can be preserved at −196 °C for several years without losing their viability. Cryopreservation of animal gametes is a component in the strategy to save endangered species by facilitating the storage of their gametes in the gene bank. It is a tool for long-term storage of spermatozoa. It was evolved as an attempt following the general short-term preservation of gametes at low temperature ranging from 0 °C to 4 °C.

The major focus had been the standardization of the protocol and practical application of the cryopreserved spermatozoa of fish in artificial fertilization and seed production. The protocol has been standardized for the cryopreservation of nearly 200 freshwater and 40 marine species spermatozoa. In India too, cryopreservation protocols have been developed for some freshwater carps, catfish, mahseer, and coldwater snowtrout.

Cryopreservation has been done for rainbow trout, brown trout, Atlantic salmon, Tilapia, Asian seabass, Atlantic halibut, Milk fish, Black porgy and Bluefin tuna, cat fishes, Walleye, Whitefish, Striped bass, Grey mullet, Common carp, Sutchi catfish, Grass carp, Rohu, Grouper, Muskellunge, *Oryzias latipes,* and Yellowtail flounder. Successful cryopreservation attempts have also been made for commercial as well as endangered species such as Beluga, Starlet, Siberian sturgeon, European sturgeon, Russian sturgeon and Shortnose sturgeon, Turbot, Gilthead seabream, European seabass, European and Japanese eel, Striped bass, Sharpsnout seabream, Halibut, Winter flounder, Dusky and Malabar grouper, Cod and Haddock.

C. J. Betsy, C. Siva, *Fisheries Biotechnology and Bioinformatics*,
https://doi.org/10.1007/978-981-99-6991-3_13

13.1 Diluents

Diluents are essential for any cryopreservation process. The mixture of extenders and cryoprotectants at various ratios is called diluents. Raw or neat milt is not suitable for cryopreservation which must be diluted with diluents (Bozkurt et al. 2012).

13.1.1 Extenders

Salt concentration present in the milt keeps it highly viscous in a liquid state. This high viscosity prevents temperature exchange and causes cryoinjuries. Hence the milt cannot be frozen directly and it has to de diluted before freezing. The dilution must be made with a solution that will not harm the spermatozoa that has to be frozen. Also the solution must have salt concentrations similar to that of the seminal plasma of milt. Such a solution is called as "Extender."

An extender is a solution that contains organic and inorganic chemicals which increase the efficacy of cryopreservation of sperm. Extender can be defined as a solution of salts, sometimes including organic compounds, which helps maintain the viability of cell during refrigeration (Graybill and Horton 1969). Extenders used for diluting the milt must be compatible with the physicochemical composition of the seminal fluid of the fish species. A successful extender must have osmolality similar to that of seminal fluid.

It was reported that extender was first developed in animals for cryopreservation of bull sperm. The extender used for bull sperm cryopreservation had egg yolk as a co-cryoprotectant. Lipids present in egg yolk protected bull sperm from "cold shock" and the sensitivity of cells to reduced temperatures. When glycerol was combined with lipid, it formed the base for the discovery of Tris-egg yolk-glycerol for freezing bull sperm, which has now become a standard.

Osmolality of seminal plasma plays an important role in spermatozoa activation. Extenders are developed only based on the osmolality of seminal plasma (Ohta and Izawa 1996). Osmolality of the extenders is adjusted by the addition of Na^+ and K^+ levels. Sperm quiescence in undiluted semen occurs in the range of 270–300 m Osm/kg. Asturiano et al. (2004) reported that the seminal plasma osmolality in the European eel is in the range of 325–330 m Osm/kg and hence extenders having osmolality in this range will suppress the motility of spermatozoa.

13.1.1.1 Types of Extenders

Extenders are of two types, namely

- Sugar-based extenders
- Ionic extenders or seminal plasma mimicking medium

Sugar-Based Extenders

These extenders are made up of sugars like glucose and sucrose. Sugars can alter the physical properties of the freezing media and also lead to better survival during freezing and thawing. These extenders can also play the role of external cryoprotectants and membrane stabilizers.

Ionic Extenders

Ionic extenders generally mimic the seminal fluid of the particular fish species. This type of extender can increase insemination volume and increased fertilization if storage times are short. Saline solution that is normally referred as physiological saline can also generate the same effect (Scott and Baynes 1980; Stoss 1983).

13.1.1.2 Role of Different Salts Used in Extenders

Extenders comprise of different organic and inorganic salts which have their own functions in protecting the spermatozoa during cryopreservation. Many of the extenders used for cryopreservation of fish milt consist of NaCl, KCl, $CaCl_2$, $MgCl_2$, and $NaHCO_3$ at different concentrations. Agarwal (2011) listed the salts and their possible roles in the extender used for cryopreservation of spermatozoa.

- $NaHCO_3$, $KHCO_3$, and TRIS—act as a buffer
- NaCl—maintains the tonicity
- KCl—prevents the activation of sperm
- Mg, Ca ions—normal components of seminal fluid
- Sugars—form a source of energy
- Glycine—improves the survival of sperm
- Mannitol—protects the sperm against toxic effect of DMSO
- Lecithin, egg yolk, promine D—protect the sperm membrane

13.1.1.3 Properties of Extender

An ideal extender for sperm should have the following properties Graham (1978):

- It may contain antibiotics that can inhibit bacterial contamination of the sperm fluid.
- It must allow substantial increase of sperm volume to maximize insemination potential.
- It must contain constituents with metal-complexing properties that can effectively bind heavy metals which are harmful to spermatozoa.
- It must have a pH of between 6 and 8 to provide buffering capacity against deleterious effects of shifts in the pH as metabolic wastes accumulate.
- It should be stable to resist enzymatic and non-enzymatic degradation.
- It should maintain proper electrolytic balance and osmotic pressure within the cells.
- It should not give great salt impacts on the cells.
- It should provide an environment suitable for the normal metabolic activities of the spermatozoa.

13.1.1.4 Functions of Extender

An extender is known to increase the efficacy of the sperm by optimizing osmotic pressure and pH, providing an energy source to prevent undesirable use of intracellular sperm phospholipid, preventing bacterial contamination with an antibiotic, and allowing semen dilution while offsetting the deleterious effect on survival produced by high dilution.

13.1.2 Cryoprotectants

During cryopreservation, the cells are subjected to cold shock and cryoinjuries. Substances that are used to protect the cells from cryoinjuries are called cryoprotectants. The principle by which the cryoprotectants work in the cells is explained by Scott and Baynes (1980). The cryoprotectants added to the diluents can bind water molecules and reduce pure ice crystal formation in the cell when they are cryopreserved. Cryoprotectants also protect the sperm cell from cold and hot shocks and prevent cell dehydration. Enzymes in the cells can also be protected, and proteins are stabilized in unfrozen and aqueous solutions. Several cryoprotectants such as glycerol, DMSO, methanol, and DMA have been used as cryoprotectant for fish sperm. The selection of cryoprotectant for a particular sperm is also to be done after detailed standardization. This is because of the variation in the makeup of the sperm fluid. Hence there is no single cryoprotectant that can be identified as universal cryoprotectant.

Cryoprotectants are chemicals that dissolve in water and lower the melting point of water. During the cryopreservation process, the water present inside the spermatozoa or any somatic cells will tend to form ice crystals. This formation of ice will eventually damage the organelle present inside the cell and will lead to cell death. During exposure to low temperature, the formation of ice outside the cells make smaller pockets of unfrozen liquid. Due to the presence of cryoprotectants in the medium, the pockets thus formed will be larger in size. Larger unfrozen pockets of cells reduce damage from freezing injury, mechanical damage from ice, and excessive concentration of salt.

13.1.2.1 Chemicals with Cryoprotective Properties

Some of the chemicals have cryoprotective properties and hence they can be used as cryoprotectants during the process of cryopreservation. They include acetamide, L-alanine, ammonium acetate, sodium iodide, chloroform, sorbitol, choline, dextran, sodium chloride, diethylene glycol, pyridine- N-oxide, glycerol-glycerophosphate, dimethyl acetamide, dimethyl formamide, dimethyl sulfoxide, hydroxyethyl starch, trimethylamine acetate, glycerol, sodium sulfate, polyvinylpyrrolidone, mannitol, ribose, ethanol, inositol, lactose, sodium bromide, polyethylene glycol, methyl urea, ethylene glycol, glycine, scrine, magnesium chloride, erythritol, magnesium sulfate, glycerol monoacetate, maltose, phenol, sodium nitrate, mannose, pluronic polyols, triethylene glycol, sucrose, proline, formamide, urea, methanol, methyl acetamide, trehalose, methyl formamide, glucose, valine, xylose, propylene glycol.

13.1.2.2 Types of Cryoprotectants

Cryoprotectants can be classified into two major types, viz.

- Penetrating or permeating cryoprotectants
- Non-penetrating or non-permeating cryoprotectants

Penetrating Cryoprotectants

They are relatively small molecules which can penetrate into the cells through cell membranes. The penetrating cryoprotectants are the commonly used cryoprotectants. The molecular mass of penetrating cryoprotectants is typically less than 100 daltons. The role of penetrating cryoprotectants is reducing ice formation and preventing cell dehydration during freezing. During vitrification, it completely prevents ice formation. Penetrating cryoprotectants include glycerol, DMSO, methanol, etc.

Glycerol

Glycerol is an organic chemical that has cryoprotection property. In 1949, a team of researchers involving Polge, Smith, and Parkes in UK identified the cryoprotective use of glycerol. In nature, it is a radio protective agent, and it is highly polar. It can interact with water through hydrogen bonding. It helps to shift the osmotic balance and maintain some water in the cells, which is very important in reducing the dehydration effects of freezing. It acts by adding to the colligative properties of a solution and reducing the salt concentration during freezing. It was the cryoprotectant used in the first attempt of cryopreservation of fish spermatozoa by Blaxter during 1953. In general, it is considered as an effective cryoprotectant for many fish species including grey mullet, white fish, and common carp.

Dimethyl Sulfoxide

It is an inorganic chemical came into use as a cryoprotective agent during 1960s. Dr. Stanley W. Jacob is considered as the "Father of Dimethyl Sulfoxide (DMSO)." It has been found that DMSO was superior to glycerol for protecting erythrocytes during freezing. DMSO is a widely used cryoprotectant for cyprinids. However, it can be effectively used for salmonids and other freshwater fishes too. Although DMSO is fairly permeable to membranes, its toxicity can be minimized by reducing the temperature. It is also a radio protective agent. Due to its gene activation properties, there may be a possibility of genetic alteration during cryopreservation (Ashwood-Smith 1986). Using DMSO at a concentration of 10% has been advocated for cyprinid sperm (Lahnsteiner et al. 2003).

Methanol

Methanol is also used as a penetrating cryoprotectant in cryopreservation of many fish species. Methanol penetrates and leaves cells rapidly and therefore, there can be a reduced equilibration time. Methanol increases the post-thaw motility rate and fertilization rate as seen in Ictalurid, silurid, and acipenserid semen. Methanol was reported as the best cryoprotectant for the cryopreservation of fish species like zebra fish, salmonids, catfishes, and tilapia.

Dimethyl Acetamide

It can be used as an alternative to other common cryoprotectants like glycerol and DMSO. In general, it is not widely used as a cryoprotectant in fishes. However, there are some recommendations for its use as a cryoprotectant because DMA can help in the post-thaw activation of spermatozoa and there is zero latency in motility activation of *Bryconorthotaenia* (Melo and Godinho 2006). Yasui et al. (2009) reported increased post-thaw motility when DMA was used at 10% concentration for the cryopreservation of loach spermatozoa.

Propylene Glycol

It produces less osmotic shock due to their rapid penetration into cells. In the cryopreservation of spermatozoa of yellowtail flounder and Atlantic halibut, propylene glycol has been identified as an effective cryoprotectant (Billard et al. 1993; Richardson et al. 1995). It is less toxic than other cryoprotectants.

Non-penetrating Cryoprotectants

They are large molecules, usually polymers, added to diluting solutions. They inhibit ice growth by the same mechanism as penetrating cryoprotectants do. But they do not enter into the cells. They are usually less toxic than penetrating cryoprotectants at the same concentration. They are also used to increase the tonicity (osmotically active concentration) of vitrification solutions, thus preventing injuries during chilling.

These compounds can form extensive hydrogen bonds with water, thus reducing water activity to a greater extent. These cryoprotectants are sometimes referred as "co-cryoprotectants." Non-penetrating cryoprotectants include protein like milk, egg yolk, bovine serum albumin; sugars such as glucose, sucrose; synthetic polymers like polyethylene glycol (PEG) and polyvinylpyrrolidone (PVP). There are numerous benefits due to the use of co-cryoprotectants. However, the nature of such co-cryoprotectants for use in cryopreservation is a matter of concern.

Egg Yolk

Egg yolk is a non-permeating cryoprotectant, and it does not penetrate through the plasma membrane of a living cell. Therefore, it can make difference in the osmotic pressure of the extender solution. Egg yolk has been tested as an additive material to extenders and cryoprotectants in many cryopreservation studies. The presence of fresh hen egg yolk in the freezing medium can protect the cell membranes during the freezing processes (Christensen and Tiersch 2005). Alvarez and Storey (1993) stated that addition of egg yolk proteins to the freezing medium aids in the recovery of sperm motility. The low density lipoprotein fraction (LDL) present in egg yolk associates with cell membranes and provides protection against injury during the cryopreservation process (Babiak et al. 2001).

Glucose

Fish spermatozoa are capable of utilizing extracellular carbohydrates like fructose, galactose, and glucose through oxidation (Gregory 1968). Glucose provides membrane protection to spermatozoa and serves as an external cryoprotectant (Maisse 1996). The presence of the carbohydrates, fructose, galactose, and glucose, although conflicting with the results reported by Cruea (1969) for *C. carpio*, indicates alternative or secondary sources of energy for sperm motility and survival in *C. carpio* and *O. mossambicus* (Kruger et al. 1984). Bozkurt et al. (2009) reported a longer duration of motility in grass carp spermatozoa cryopreserved with an extender containing an external energy source of 300 mM glucose. Sugars alter the physical properties of the freezing media and allow for better cell survival during freezing and thawing. Stoss and Refstie (1983) demonstrated that glucose provides excellent cryoprotection in the absence of other extender components and without additional equilibration.

Glucose is also a non-permeating cryoprotectant. It is otherwise a substance used for providing or deriving energy to fish semen while in most other animals, it is fructose. There are evidences that addition of glucose to the extender can enhance the quality of milt and the storage duration.

Bovine Serum Albumin

Bovine Serum Albumin (BSA) is a serum albumin protein derived from cows. It is often used as a protein concentration standard. It has 583 amino acid residues with a molecular weight of 66.5 kDa. It has been widely used for cattle and human sperm preservation. BSA eliminates free radicals generated by oxidative stress and protects membrane integrity of equine and human sperm cells from heat shock during freeze-thaw process of semen. BSA can induce increased motility and viability of spermatozoa after long-term storage in low temperatures and cryopreservation. It has identified to stimulate sperm motility by an unknown mechanism in various mammalian species.

13.2 Dilution

Milt dilution ratio is very important for fish sperm to survive after cryopreservation (McPartlin et al. 2008). Lahnsteiner (2000) proposed that cell concentration in the extenders should not exceed 2.0×10^9 cells/mL to 3.0×10^9 cells/mL of extender when sperm is frozen in straws. At higher cell concentrations, the post-thaw fertility of milt significantly decreases, probably as a result of cell compression because of limited intracellular space. Hence dilution of milt is essential which depends on the constituents of the extender, concentration of the spermatozoa in gonad, the sperm physiology of the species, and on the freezing protocols (Sarder et al. 2011).

Billard and Cosson (1992) stated that because of high sperm concentration in milt, assessment of motility requires high dilution rate. Dilution is known to reduce sperm motility by removing protective components of seminal plasma in rainbow trout. Three parts diluents to one part milt are commonly used for fish sperm with greater success (Scott and Baynes 1980). Semen dilution is a process that has been carried out as a means to increase the number of eggs that can be fertilized with a small volume of semen (Billard 1985). The motility and fertility of deep frozen spermatozoa of *C. carpio* were reported to be significantly improved when the dilution ratio was reduced from 1:100 to 1:2 (Cognie et al. 1989). According to Rana and McAndrew (1989) and Gwo et al. (1991), the milt to dilution ratio had strong effect on post-thaw motility and fertility. Erdahl and Graham (1987) mentioned about a concept called "dilution effect" which causes sperm cells to become motile at some point during storage and decrease fertility. Stoss (1983) stated that this dilution effect may be related to the capacity of sperm cells to re-establish ionic equilibrium after ionic changes caused by the dilution medium.

13.3 Process of Cryopreservation

The protocol of cryopreservation varies considerably for individual species. This variation may be due to difference in the spermatological parameters which determine the use of specific extender, cryoprotectant, dilution ratio, and equilibration time. Till date there is no single common protocol available for both freshwater and marine fishes which can be followed for cryopreservation of all species. Parameters which influence the cryopreservation method include the type and concentration of extender and cryoprotectant, dilution ratio, equilibration time, rate of cooling, method of freezing like straw or pellet, and thawing rate. Standardization of protocol needs many trials and attempts with different combinations and comparisons of success rates. The success of cryopreservation can be evaluated or determined by the fertilization rate.

The process starts from brooder selection and ends with freezing of milt loaded straws in liquid nitrogen. Care must be taken in every step to ensure success. The following are the steps involved in the process of cryopreservation.

13.3.1 Selection of Brooders

The cryopreservation process of fish spermatozoa starts from the identification of suitable donors. The male brooders which are matured should be selected based on the visual assessment of the oozing milt and normal flow of milt for a gentle squeeze of the belly. The maturity of fish can be assessed by gentle pressing of the belly region on which milt must ooze out freely. The brooders can be obtained either from hatchery or natural water bodies.

13.3.2 Milt Collection

Once suitable donors are selected, milt is collected from them. Appropriate fish sperm collection methodology is essential for the successful preservation of fish

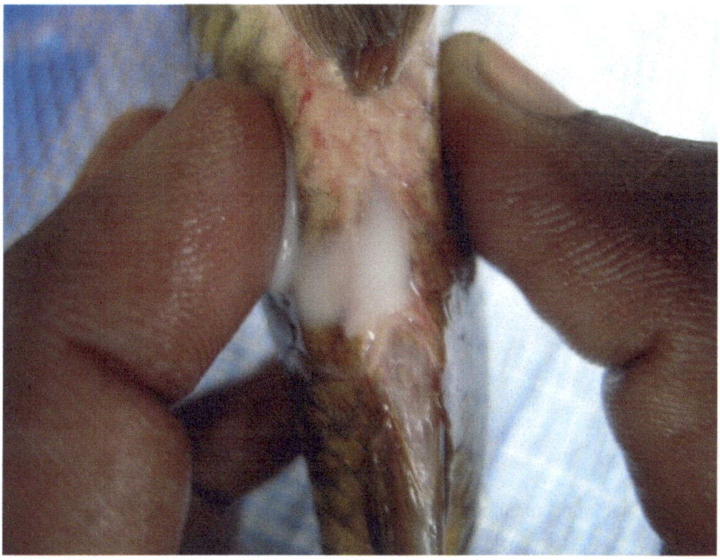

Fig. 13.1 Milt oozing out from belly

spermatozoa. Rao (1989) was of the opinion that a suitable fish sperm collection method would reduce the internal and external injuries to the donor fish. While describing the method, Horvath et al. (2003) stated that prior to collection of milt, the abdomen of the donor should be wiped off carefully to remove water and then gentle pressure should be applied in the abdomen area to expel urine. After the expulsion of urine, the milt can be collected either by massaging the belly or using catheter (Fig. 13.1). Lubzens et al. (1997) stripped the gametes after careful drying of genital pores in order to avoid any contamination of gametes with urine or water.

Milt can be collected from brooders after hormonal injection (Fig. 13.2). Before collection, the genital pore region must be cleaned with absorbent cotton and double ply tissue paper to remove any moisture or mucus in that area. Urine must be evacuated by mild gentle press near the genital pore anteriorly, that could also press out the fecal matter if any. Bleeding or traces of blood in the milt is considered unsuitable and those donors and collection should be culled out.

After a complete wiping and cleaning of the genital pore, gentle pressure must be applied in the ventral portion nearer to the pelvic fins and the fingers must be moved from anterior to posterior part of the body to evacuate the milt present in the tubular testes. This results in the oozing of milt through the genital pore which can be collected in a pre-labeled, sterile cryovial (Fig. 13.3). Milt oozing out during the first three to four stripping alone should be collected as the subsequent stripping may lead to bleeding or urine contamination.

Milt can also be collected using catheter and syringes. In this method, catheter or 2.5 mL needle-less syringe is inserted into the genital pore area and gentle pressure is given in the abdomen region to collect the milt. Care should be taken to prevent

Fig. 13.2 Hormonal inducement of male carp

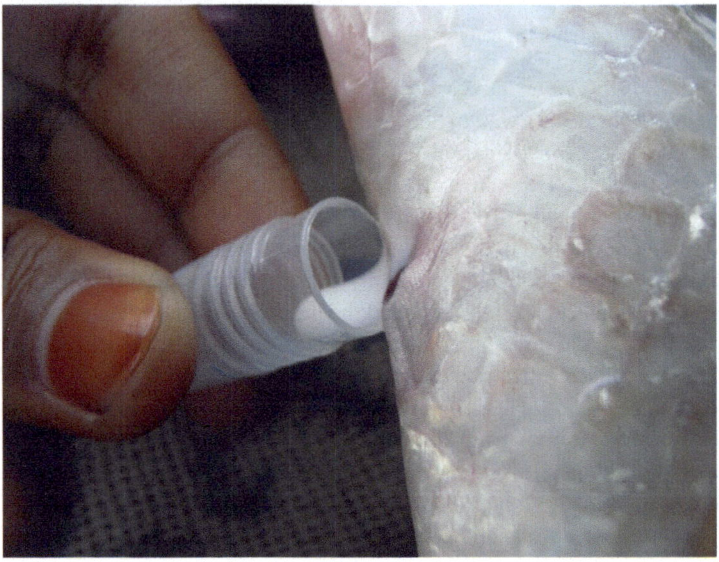

Fig. 13.3 Milt being collected in a cryovial

contact of the tip of catheter or syringe with water, mucus, or urine. Babiak et al. (1998) used syringes for collecting milt in Asp (*Aspius aspius*). Jing et al. (2009) collected milt from zebrafish by three methods like dissecting the testis without crushing, dissecting the testis with crushing and abdominal massage. They reported higher sperm motility when milt was collected by dissecting the testis without crushing and abdominal massage.

Use of catheter resulted in more consistent osmolality (300–330 m Osmol/kg) in comparison with samples collected by abdominal massage (180–290 m Osmol/kg). Graduated cryovials are used for milt collection. Individual vial should be used for individual donors so as to avoid unexpected contamination leading to wastage of the complete collection. All the vials must be carefully sealed and kept in the icebox with gel ice at 10–15 °C to avoid temperature fluctuations and photo shocks.

13.3.3 Diluent Preparation

13.3.3.1 Dilution Ratio

Dilution increases the volume of milt. There is no fixed dilution ratio for any species. Milt is generally diluted by 3–100 folds based on density and viscosity of the seminal plasma. According to Rana (1995), milt of salmonid, carp, and tilapia were diluted to 3–20 folds. Before fixing the dilution ratio, optimal sperm to egg ratio must be determined. After fixing up of dilution ratio, the diluents must be slowly added to the milt in order to avoid changes in the pH and osmolality of the sperm. The sample should be checked for absence of motility immediately after addition of the diluent.

13.3.4 Equilibration of Milt

The period between the addition of diluents with the milt and the freezing of milt is the equilibration time. The equilibration time allowed for the sperm should be minimal due to the small size of spermatozoa. Most of the cryopreservation process does not provide equilibration time. Billard (1978) stated that when equilibration exceeds 2 min, fertilizing ability drops. Stoss and Refstie (1983) and Jamieson (1991) reported that since spermatozoa are small, the penetration of cryoprotectants is rapid and no equilibration time is required and hence the toxic effect of DMSO can be minimized. Harvey (1983) stated that the detrimental effects of any equilibration time were aggravated by increasing the concentration of cryoprotective agent. However, Ogier de Baulny (1997) observed that DMSO needed 10 min to penetrate into rainbow trout spermatozoa although the fertilizing capacity of frozen-thawed spermatozoa was same with or without equilibration period.

13.3.5 Packing of Milt

The equilibrated samples are frozen after packing it in appropriate methods. This can be done by three ways, viz. pellet method, vial method, and straw method. In pellet method, the equilibrated samples are formed into pellets in small hole drilled into solid dry ice (−79 °C). French straws are the plastic straws of various volumes like 0.25 mL and 0.5 mL which are available in different colors to differentiate between samples. This color variation greatly helps in the storing of milt of different species in a single canister minimizing the confusion. One end of the straw is cotton sealed, and the other is open ended.

The filled straws are sealed with a polymer powder and dipped in cold water for proper sealing. This powder is used to seal the open end of straws while the other end is already sealed with cotton. When the straws are loaded with milt because of the cotton seal and also the filled condition, the fluid will not come out of the straws. Still when the straws are removed from the loader, there is a possibility that the fluid may come out of the straws. Therefore, on completion of loading the straws, the open end of the straws will be dipped in sealing powder and consequently in cold water for solidification of the polymer. This will make the open end of the straws being sealed. After sealing, the sealed straws must be taken from water and wiped with tissue paper to remove water droplets and moisture from the straws before putting them for freezing.

13.3.6 Pre-freezing

Cooling and thawing rates are regarded as the most critical phases of cryopreservation (Franks 1985). During cooling, the cells may be subjected to various stresses. Due to increased solute concentration in the residual solution, water is converted to pure solid during freezing and is generally regarded as the single most important stress factor for cells (Franks 1982; Taylor 1984, 1987). Under rapid cooling conditions, the rate of water flow out of the cell may not be fast enough and the cell equilibrate by freezing internally. Under these conditions, stresses caused by intracellular ice crystallization and freeze concentration of cytoplasm may be significant (Franks 1982). Following ultra-rapid cooling, however, the intracellular ice crystals will be very small and thermodynamically unstable (Taylor 1987).

The packed milt must be pre-frozen before final freezing with liquid nitrogen. This can be done by arranging the vials or straws horizontally on aluminum rack and by keeping it above liquid nitrogen (LN_2) vapor in a thermocol box covered by lid for rapid freezing. In order to get uniform freezing effect, the distance between the vials/straws and LN_2 must be maintained at 5 cm. This can be done at different rates of temperature and time like 10–45 °C per min until the temperature reaches −70 °C. Pre-freezing is done for 10–15 min depending on the volume of the sample and dilution ratio adapted. It can be done by using programmable freezers also.

Sultana et al. (2010) used a computer-controlled freezer (CL 3300, Australia) to freeze the milt samples of common carp. They followed a two-step freezing protocol in which firstly the milt was cooled from 5 °C to −4 °C at a rate of 4 °C per minute (Ramp 1 min), then −4 °C to −80 °C at a rate of 10 °C per min and held for 10 min. Afterwards the straws were taken out from the cryochamber and plunged into the liquid nitrogen (−196 °C). Whenever needed, straws were retrieved from the liquid nitrogen and thawed in water bath at 40 °C for 7 s.

Similarly, Sarder et al. (2011) used computer controlled-rate freezer (CL-3300, equipped with the software Cryogenesis, version 4, for Windows, Cryologic, Pty Ltd., Australia, 1998 and 1999) to freeze rohu milt. Two-step freezing protocol was used where sample was first cooled to −4 °C at a temperature decreasing rate of 4 °C/min, and then to −80 °C at 10 °C/min. After freezing samples were removed from the cryochamber and loaded in the canisters and placed into the liquid nitrogen (−196 °C) for 24–48 h. The post-thaw motility of the sperm was checked by thawing the sperm at room temperature (25–26 °C) for about 1 min.

13.3.7 Freezing and Storage

During freezing, all the biological activities of the cell stop. Freezing is done at a two-step procedure, where milt is cooled in nitrogen vapor on a floating tray or in the neck of a container and then the straws are plunged into liquid nitrogen (Suquet et al. 2000). Billard et al. (2004) used programmable freezer for freezing sturgeon sperm. The regime they followed was 3.5–5 °C/min from +2 to −14 °C, 15–20 °C/min down to −70 °C and direct transfer to liquid nitrogen.

Chao et al. (1987) had frozen the straws of Tilapia (*Oreochromis* sp) milt at a rate of 5 °C/min to the final temperature of −75 °C using dry ice. They observed better motility at a rate of −35 °C and suggested a freezing rate of 1.54 °C/s when programmable freezer is used. Scott and Baynes (1980) opined an optimum cooling rate for freezing salmonid spermatozoa to be in the range of 30–160 °C/min. Baynes and Scott (1987) followed a cooling rate of 80 °C/min over the phase from 0 °C to −60 °C, during cryopreservation of Rainbow trout (*Salmo gairdneri*) spermatozoa.

If the cell is sufficiently small, not exceeding a few microns, as in microorganisms, so much water can be withdrawn during freezing that they are effectively desiccated (Luyet 1938). It is possible to predict such optimum cooling rates using mathematical models provided that values of factors required for the models are known. These factors are as follows:

- The viscosity of the intracellular solution.
- The surface to volume ratio of the cell.
- The rate of diffusion of water through the cell membrane.
- The distance between the cell and the nearest ice crystal.
- The viscosity of the extracellular solution.
- The permeability of the added cryoprotectants.

The pre-frozen milt samples must be transferred to canisters containing liquid nitrogen in a Cryocan and are stored at −196 °C. Each canister must be labeled with the details of the species name and date of cryopreservation.

13.3.8 Thawing

Morris (1981) stated that thawing temperature and duration are critical factors in the survival of cryopreserved sperm cells. Thawing is the process for the revival of cryopreserved spermatozoa. Sperm samples are usually thawed in a temperature-controlled water bath or in air. Thawing temperature plays an important role in the post-freeze semen quality of fish. Thawing must be conducted quickly as that of freezing to prevent cell injury (Stoss 1983). Fast thawing temperature decreases the recrystallization effect in spermatic cells and hence reduces the membrane damage. Lahnsteiner (2000) reported that thawing rates should be high to avoid recrystallization. Bozkurt et al. (2012) stated that high temperature is needed to recover membrane stability and metabolism of spermatozoa.

Different temperature has been used to thaw cyprinid semen like 40 °C (*C. carpio* L.), 30 °C (*C. carpio*), and 25 °C (*C. chalcoides*) (Babiak et al. 1995; Magyary et al. 1996; Lahnsteiner et al. 2000). Lahnsteiner et al. (2000) observed the optimal thawing temperature of *C. carpio* was 25 °C within a time of 15–45 s. If the freezing of spermatozoa is done in a rapid rate, the thawing should also be done at the same rate in order to prevent cell injury by water recrystallization.

Post-thaw motility is the most important parameter which determines the success of cryopreservation protocol. Thawed vials or straws are wiped with tissue to remove water, the straws are cut open, and the samples are poured into an empty vial for easy analysis. The motility duration and percentage of motile sperm must be estimated. The success of cryopreservation depends on the recovery of post-thaw motility rate and its ability to fertilize the eggs.

13.4 Cryopreservation in Aquaculture

Although cryopreservation is not a recent happening, in aquaculture it is found adopted only in recent years. Advent of artificial insemination has brought in revolutionary changes in the dairy industry. Cryopreservation of spermatozoa at sub-zero temperature has been considered as a viable technique for the success of organized breeding programmes through artificial insemination. Introduction of this biotechnological approach in fish breeding programmes becomes necessary to ensure successful breeding and continuous seed production even during monsoon failures. While induced breeding helps in one aspect by making the female brooders to evaluate and produce eggs, the cryopreserved male gametes can possibly help in the non-use of male brooders. This technology has received due attention in recent years for the improvement of offspring quality and availability of fish fry for stocking throughout the year. Cryopreservation and application of cryopreserved spermatozoa for the breeding of fishes may be a possible answer to produce quality seeds and genetically improved varieties overcoming the adversely affecting natural monsoon conditions.

Cryopreservation of gametes has been done to increase the longevity of gametes for several years without any drastic changes in their efficiency or potency. In aquaculture, it principally focuses on cross breeding of fishes within the populations and aids in saving endangered species by facilitating the storage of their gametes in the gene bank. Production of hybrids with desired characters is a possible major advantage in the application of cryopreservation in aquaculture. Cryopreservation could offer benefits like synchronization of gamete availability of sexes, sperm economy, simplification of broodstock management, transport of gametes from different fish farms, and germplasm storage for genetic selection programs or conservation of species. Aquaculture industry can be developed through these benefits.

Sperm cryopreservation can lead to the germ storage of all transgenic lines. Since DNA damage may impair fertility or embryo development, studies on DNA have been made using approaches like comet assay (single cell gel electrophoresis), TUNEL (terminal deoxynucleotidyl transferase-nick-end-labeling), SCSA (sperm chromatin structure assay), and the analysis of specific DNA sequences using RT-PCR. In addition to being a simple technique for the preservation of gametes, cryopreservation can contribute immensely for the gene mapping studies in aquaculture and subsequent gene manipulation.

References

Agarwal NK (2011) Cryopreservation of fish semen. In: Thapliyal M, Thapliyal A, Bhatt JP (eds) Himalayan aquatic biodiversity conservation & new tools in biotechnology, vol 2011. Transmedia Publication, pp 104–127

Alvarez JG, Storey BT (1993) Evidence that membrane stress contributes more than lipid peroxidation to sublethal cryodamage in cryopreserved human sperm: glycerol and other polyols as sole cryoprotectant. J Androl 14(3):199–209

Ashwood-Smith MJ (1986) The cryopreservation of human embryos. Humanit Rep 1:319–332

Asturiano JF, Pérez L, Garzón DL, Marco-Jiménez F, Peñaranda DS, Vicente JS, Jover M (2004) Physio-chemical characteristics of seminal plasma and development of media and methods for the cryopreservation of European eel sperm. Fish Physiol Biochem 30:283–293

Babiak I, Glogowsky J, Brzuska E, Szumiec J, Adamek J (1995) Cryopreservation of sperm of common carp, Cyprinus carpio. Aquac Res 28:567–571

Babiak I, Glogowski J, Kujawa R, Kucharezyk D, Mamearz L (1998) Cryopreservation of sperm from Asp Aspius aspius. Prog Fish Cult 60:146–148

Babiak I, Glogowski J, Goryczko K, Dobosz S, Kuzminski H, Strzezek J, Demianowicz W (2001) Effect of extender composition and equilibration time on fertilization ability and enzymatic activity of rainbow trout cryopreserved spermatozoa. Therio 56:177–192

Baynes SM, Scott AP (1987) Cryopreservation of rainbow trout spermatozoa: the influence of sperm quality, egg quality and extender composition on post thaw fertility. Aquaculture 66:53–67

Billard R (1978) Changes in structure and fertilizing ability of marine and freshwater fish spermatozoa diluted in media of various salinities. Aquaculture 14(3):187–198

Billard R (1985) Artificial insemination in salmonids. In: Iwamotoa RN, Sewer S (eds) Salmonid reproduction. Washington Sea Grants Publication, Seattle, pp 116–128

Billard R, Cosson MP (1992) Some problems related to the assessment of sperm motility in freshwater fish. J Exp Zool 261:122–131

Billard R, Cosson J, Crim LW (1993) Motility of fresh and aged halibut sperm. Aquat Living Resour 6:67–75

Billard R, Cosson J, Noveiri SB, Pourkazemi M (2004) Cryopreservation and short-term storage of sturgeon sperm-a review. Aquaculture 236:1–9

Bozkurt Y, Ogretmen F, Secer FS (2009) Effect of different extenders and storage periods on motility and fertilization success of grass carp (*Ctenopharyngodon idella*) sperm during spawning season. Tarim Bilimleri Dergisi 15(3):277–284

Bozkurt Y, Yavas I, Karaca F (2012) Cryopreservation of brown trout (*Salmo trutta macrostigma*) and ornamental koi carp (*Cyprinus carpio*) sperm. In: Katkov I (ed) Current frontiers in cryopreservation. InTech Publishers, pp 293–304

Chao NH, Chan WC, Liu KC, Liao IC (1987) The properties of Tilapia sperm and its cryopreservation. J Fish Biol 30:107–119

Christensen MJ, Tiersch RT (2005) Cryopreservation of channel catfish sperm: effects of cryoprotectant exposure time, cooling rate, thawing conditions and male-to-male variation. Therio 63: 2103–2112

Cognie F, Billard R, Chao NH (1989) Freezing of the milt of the common carp, Cyprinus carpio. J Appl Ich 5:165–176

Cruea DD (1969) Some chemical and physical characteristics of fish sperm. Trans Am Fish Soc 98: 785–788

Erdahl AW, Graham EF (1987) Fertility of teleosts semen as affected by dilution and storage in a seminal plasma mimicking medium. Aquaculture 60:311–321

Franks F (1982) Physiological water stress. In: Franks F, Matthias SF (eds) Biophysics of water. John Wiley and Sons, Chichester, pp 279–294

Franks F (1985) Biophysics and biochemistry at low temperatures. Cambridge University Press, Cambridge

Graham EF (1978) Fundamentals of the preservation of spermatozoa. In: Grisamore NT, Sieber JM (eds) The integrity of frozen spermatozoa, Proceedings of a round-table conference, Washington, pp 4–44

Graybill JR, Horton HF (1969) Limited fertilisation of steelhead trout eggs with cryopreserved sperm. J Fish Res Bd Canada 26:1400–1404

Gregory RW (1968) Occurrence of fructose in trout seminal plasma. Trans Am Fish Soc 97:203–204

Gwo JC, Strawn K, Longnecker MJ, Arnold CR (1991) Cryopreservation of Atlantic croaker spermatozoa. Aquaculture 94:355–375

Harvey B (1983) Cryopreservation of Sarotherodon mossambicus spermatozoa. Aquaculture 32: 313–320

Horvath A, Miskoiczi E, Urbanyi B (2003) Cryopreservation of common carp sperm. Aquat Living Resour 16:457–460

Jamieson GM (1991) Fish evolution and systematic: evidence from spermatozoa. Cambridge University Press, Cambridge, p 319

Jing R, Huang C, Bai C, Tanguay R, Dong Q (2009) Optimization of activation, collection, dilution and storage methods for zebrafish sperm. Aquaculture 290:165–171

Kruger JCDW, Smit GL, Vuren JHJV, Ferreiera JT (1984) Some chemical and physical properties of the semen of Cyprinus carpio L. and Oreochromis mossambicus (Peters). J Fish Biol 24:263–272

Lahnsteiner F, Berger B, Horvath A, Urbanyi B, Weismann T (2000) Cryopreservation of spermatozoa in cyprinid fishes. Therio 54:1477–1498

Lahnsteiner F, Berger B, Weismann T (2003) Effects of media, fertilization technique, extender straw volume and sperm to egg ratio on hatchability of cyprinid embryos, using cryopreserved semen. Therio 60:829–841

Lubzens E, Daube N, Pekarskay I, Magnus Y, Cohen A, Yuesefovich F, Feigin P (1997) Carp (Cyprinus carpio L.) spermatozoa cryobanks – strategies in research and application. Aquaculture 155:13–30

Luyet BJ (1938) The survival of plant cells immersed in liquid air. Biodynamica 33:284–285

Magyary I, Urbanyi B, Horvath L (1996) Cryopreservation of common carp (Cyprinus carpio L.) sperm II. Optimal conditions for fertilization. J Appl Ichthyol 12:117–119

Maisse G (1996) Cryopreservation of fish semen: a review. In: Proc. refrigeration sci. technol. conf., refrigeration and aquacult. Institut International du Froid, Paris, pp 443–467

McPartlin LA, Littell J, Mark E, Nelson JL, Travis AJ, Bedford-Guaus SJ (2008) A defined medium supports changes consistent with capacitation in stallion sperm, as evidenced by increases in protein tyrosine phosphorylation and high rates of acrosomal exocytosis. Therio 69:639–650

Melo FCSA, Godinho HP (2006) A protocol for cryopreservation of spermatozoa of the fish Brycon orthotaenia. Anim Reprod 3(3):380–385

Morris GJ (1981) Cryopreservation: an introduction to cryopreservation in culture collections. Institute of Terrestrial Ecology, Cambridge

Ogier de Baulny B (1997) Cryoconservation du sperme de poissons. Evaluation des dommages cellulaires. In: Amélioration de la technique de congélation. Perméabilité membranaire aux cryoprotectants. Thesis, Rennes, p 129

Ohta H, Izawa T (1996) Diluent for cool storage of the Japanese eel (Anguilla japonica) spermatozoa. Aquaculture 142:107–118

Rana KJ (1995) Preservation of gametes. In: Bromide NR, Roberts RJ (eds) Broodstock management and egg and larval quality. Blackwell Science, Cambridge, pp 53–75

Rana KJ, McAndrew BJ (1989) The viability of cryopreserved tilapia spermatozoa. Aquaculture 76: 335–345

Rao KG (1989) Cryopreservation of carp sperm. In: Das P, Jhingaran AG (eds) Fish genetics in India. Today and Tomorrow's Printers and Publishers, New Delhi, pp 193–198

Richardson GF, Crim LW, Yao Z, Short C (1995) Cryopreservation of yellowtail flounder (*Pleuronectes ferrugineus*) semen. In: Goetz FW, Thomas P (eds) Proceedings of the fifth international symposium, reproductive physiology of fish, p 136. Austin

Sarder MRI, Rahman AKMA, Samad MS, Nazrul KMS, Bhuiyan MKJ (2011) Cryopreservation of sperm of Labeo rohita (Hamilton, 1822) and its use in fertilization and hatching of eggs. Progress Agric 22(1 & 2):123–137

Scott AP, Baynes SM (1980) A review of the biology, handling and storage of salmonid spermatozoa. J Fish Biol 17:707–739

Stoss J (1983) Fish gamete preservation and spermatozoa physiology. In: Hoar WS, Randall DJ, Donaldson EM (eds) Fish physiology (IX b). Academic Press, New York, pp 305–350

Stoss J, Refstie T (1983) Short-term storage and cryopreservation of milt from Atlantic Salmon and sea trout. Aquaculture 30:229–236

Sultana M, Nahiduzzaman M, Hassan MM, Khanam MUH, Hossain MAR (2010) Fertility of cryopreserved common carp (Cyprinus carpio) spermatozoa. Univ J Zool Rajshahi Univ 28:51–55

Suquet M, Dreanno C, Fauvel C, Cosson J, Billard R (2000) Cryopreservation of sperm in marine fish. Aquac Res 31:231–243

Taylor MJ (1984) Subzero preservation and the prospect of long term storage of multi cellular tissues and organs. In: Catne RY (ed) Transportation immunology, clinical and experimental. Oxford University Press, Oxford, pp 360–390

Taylor MJ (1987) Physio-chemical principles in low temperature biology. In: Grout BWW, Morris GJ (eds) Effects of low temperature on biological systems. Edward Arnold Ltd, London, pp 3–72

Yasui GS, Lenin AR, Fujimoto T, Arai K (2009) A sperm cryopreservation protocol for the loach, Misgurnus anguillicaudatus and its applicability for other related species. Anim Reprod Sci 116: 335–345

Molecular Techniques 14

14.1 Polymerase Chain Reaction (PCR)

PCR is defined as the DNA replication in vitro. It was developed by Kary Mullis in 1983 and received Nobel Prize for chemistry in 1993. PCR is used for the amplification of gene of interest exponentially. For every cycle of heating and cooling, the amount of template is doubled and hence approximately 1 million copies of the single target DNA molecule is produced after 20 cycles of PCR. One limitation of PCR is that the border region sequences of the DNA (to be amplified) must be known in order to select the appropriate primers which anneal (attach) at its 3′ ends.

14.1.1 Requirements of a PCR Reaction

- DNA template
- Two oligonucleotide primers which are complementary to 3′ ends of target DNA strands (forward and reverse primer) (Fig. 14.1)
- 4 dNTPs (deoxynucleotide triphosphates—dATP, dTTP, dCTP, dGTP)
- Magnesium ions
- Thermostable DNA polymerase—Taq polymerase is isolated from a thermophilic bacterium, i.e., *Thermus aquaticus*
- A thermal cycler in which PCR is carried out

14.1.2 PCR Cycle

14.1.2.1 Denaturation

The double-stranded DNA is denatured by using high temperature of 92–94 °C for 15 s. Now each separated single-strand acts as a template for DNA synthesis. DNA sequences with high GC content require larger denaturation time (Fig. 14.2).

C. J. Betsy, C. Siva, *Fisheries Biotechnology and Bioinformatics*,
https://doi.org/10.1007/978-981-99-6991-3_14

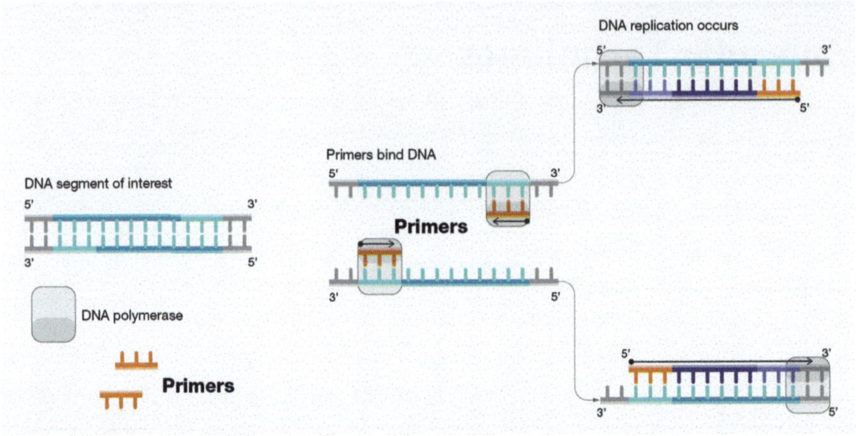

Fig. 14.1 Primers in PCR (reproduced with permission from Darryl Leja, National Human Genome Research Institute)

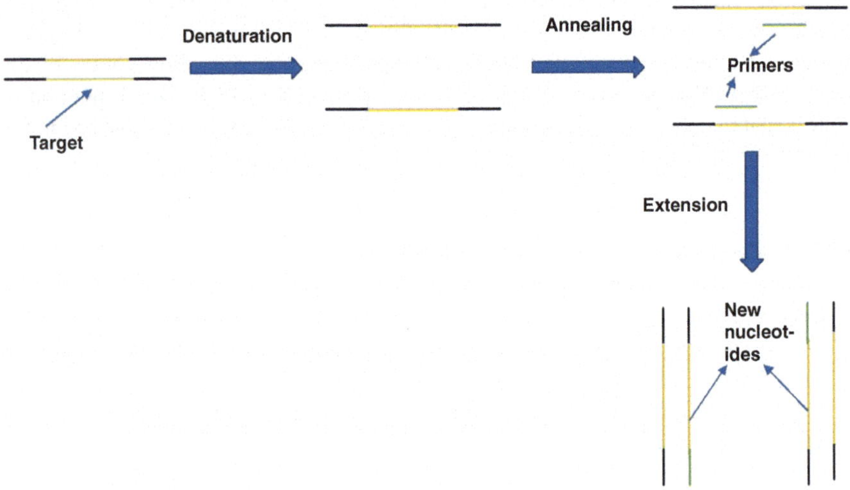

Fig. 14.2 Steps involved in PCR

14.1.2.2 Annealing

Two sets of oligonucleotide primers are used to anneal (hybridize) to synthesize complementary single-stranded target sequences. This step is carried out at a slightly lower temperature of 40–60 °C using Mg^{2+} and dNTPs (deoxynucleoside triphosphates), depending on the length and sequence of the primers.

14.1.2.3 Extension

This step is carried out at a temperature of 72 °C. Taq DNA polymerase is used in this reaction, which can tolerate the high temperature of the reaction that extends the

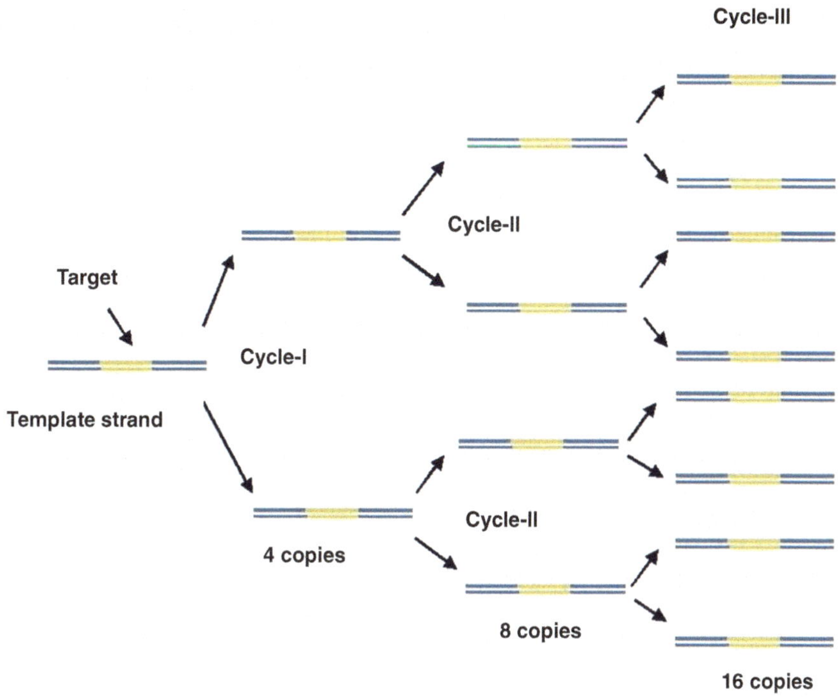

Fig. 14.3 The exponential amplification of DNA in PCR

primers by adding nucleotides complementary of the template. The extension rates of thermostable polymerases are between 2 kbp min and 4 kbp min.

The exponential amplification of DNA in a PCR cycle is depicted in Fig. 14.3.

14.1.3 Steps in PCR

- Mix target DNA sequence, excess of primers, dATP, dTTP, dCTP, dGTP, and Taq polymerase in the reaction mixture in Eppendorf tube. Place this tube in thermal cycler.
- Reaction mixture is given high temperature of about 90–98 °C for few seconds to denature the DNA. As a result, the double-stranded DNA becomes single stranded.
- Temperature is changed to about 55 °C for 20 s so that primers are annealed at 3′ ends of DNA.
- Now the temperature is maintained at 72 °C for 30 s which facilitates the functioning of Taq polymerase, thus synthesizing the complementary strand of DNA.

- Hence, 1 cycle of PCR is completed here resulting in the formation of two ds DNA molecules from one ds DNA.
- Same cycle is repeated till the required number of DNA copies are obtained.

14.1.4 Types of PCR

- Asymmetric PCR—amplifies just one strand of the target DNA.
- Colony PCR—used to screen colonies of bacteria after transformation with a plasmid.
- Hot start PCR—inactivates the Taq polymerase until the reaction starts, using antibodies that are denatured by heat.
- In situ PCR–PCR that takes place in cells, or in fixed tissue on a slide.
- Inverse PCR—amplifies DNA next to a known sequence, using primers placed in the reverse direction to normal.
- Multiplex PCR—uses a number of pairs of primers to allow analysis of a number of fragments in a single sample.
- Nested PCR—after an initial 30–35 cycles of PCR, an additional round of PCR uses new primers "nested" within the original primers, making the process more sensitive because it reduces the risk of unwanted products from primers binding to incorrect regions.
- qPCR—in quantitative PCR (also known as real-time PCR, RT PCR or qRT-PCR), the DNA or RNA molecules are tagged using fluorescent probes, so that the concentration of amplified products can be monitored and quantified in real-time by tracking the level of fluorescence,
- Repetitive sequence-based PCR–PCR using primers that are complementary to repeated non-coding sequences in the genome.
- Reverse transcriptase PCR—confusingly, also known as RT PCR—creates cDNA (complementary DNA) by reverse transcribing RNA to DNA using reverse transcriptase.

14.1.5 Uses of PCR

- To amplify DNA.
- To detect mutations and to produce in vitro mutations.
- To diagnose genetic disorders.
- To prepare DNA for sequencing.
- To analyze genetic defects in single cells from human embryos.
- To identify virus and bacteria in infectious diseases.
- To characterize genotypes.

14.2 Enzyme-Linked Immunosorbent Assay (ELISA)

ELISA is used to identify peptides, proteins, antibodies, and hormones. It is used in the fields of biotechnology and medicine as a diagnostic tool. Mainly, antibodies and color changes are used to identify target substances. Also, ELISAs are useful in measuring antigen and antibody concentration.

14.2.1 ELISA Principle

ELISA works by coupling Ab or Ag to assay enzyme. The assay combines the specificity of Ab and sensitivity of assay enzymes to primarily detect Ag through assay Ab or Ab through assay Ag.

ELISA was devised as an alternate approach for radioimmunoassays during the early 1970s. This is a plate-based assay intended toward recognition and quantification of proteins, antigens, peptides, antibodies, and hormones. The elementary enzyme-linked immunosorbent assay (ELISA), or enzyme immunoassay (EIA), is different from the other available antibody-based techniques in a way that in a polystyrene (96 or 384) multi-well plate (which is considered good for protein binding), the binding to solid surface sequentially results in the identification of specific reactions excluding the non-specific ones. This results in the generation of quantitative results. It is this obligatory binding and immobilization of components that is unique to this process making ELISAs exclusive and easy to implement.

The participating reactants in ELISA that is immobilized to the surface of the microplate enable differentiation of the reactants that have bonded out of the non-bound unspecific reactants throughout the analysis. In an ELISA, the Ag or Ab is bound onto a solid surface usually onto a microwell plate and thereafter it is interacted with a counterpart Ag or Ab. The counterpart Ag or Ab is connected to a reporter molecule like an enzyme or alternatively can be detected by a secondary antibody that is in turn linked via conjugation to an enzyme.

The detection of the assay is carried out by evaluating the enzyme activity of the conjugate which is implemented by incubation with a suitable substrate for the enzyme resulting in a generation of a quantifiable product. Generally, the ELISA technique results in a colored end product which absorbs at a particular wavelength and can be correlated to the quantity of analyte in question present in the sample. An extremely specific antibody–antigen interaction is the utmost critical component of the entire process.

14.2.2 Requirements for ELISA

- Purified antigen (if you want to detect or quantify antibody).
- Purified antibody (if you want to detect or quantify antigen).
- Standard solutions (positive and negative controls).
- Sample to be tested.

- Micro-titer plates: plastic trays with small wells in which the assay is done.
- Wash fluid (buffer).
- Enzyme-labeled antibody and enzyme substrate.
- ELISA reader (spectrophotometer) for quantitative measurements.

14.2.2.1 Enzyme Labels

Enzyme labels are used to detect the binding of antigen–antibody complex. It should have high specific reactivity. It should be easily coupled to antigen–antibody complex and must be stable. Enzymes used in labeling should not be normally present in the samples to be tested. Examples of enzyme labels are horseradish peroxidase, alkaline phosphatase, and glucose oxidase.

14.2.3 Stages in ELISA

1. Adsorption of either Ag or Ab to the microtiter plate.
2. Addition of the test sample and subsequent reagents.
3. Incubation of reactants (formation of Ag–Ab complex).
4. Separation of bound and free reactants by washing.
5. Binding of enzyme to the target antigen or antibody.
6. Addition of substrate (production of a visible signal).
7. Visual or spectrophotometric reading of the assay.

14.2.4 Methods in ELISA

The process begins with coating the multi-well plate with either antigen/antibody usually taking 30 min to overnight incubation which is followed by primary antibody reaction for 2 h after blocking. Then the application of secondary antibody is carried out with further incubation for 2 h, thereafter substrate preparation is generally done freshly just before the reaction and finally development which takes near about 30 min.

The components in the ELISA are immobilized which makes the process easy to perform. But as the assay requires the use of surface binding for appropriate separation, washing is carried out several times repeatedly after each step in order to ensure removal of unbound entities. The excess fluid is to be necessarily removed during the course of the reaction in the interest to prevent the unwanted dilution of the reagents that will be added in the subsequent steps. For ensuring the desired uniformity dedicated plate, washers are implemented in this process.

As ELISAs incorporate multistep reactions, it can become quite complex in case of measurement of analyte concentrations. In case of complicated samples such as blood, there is a presence of a significant amount of analytes that can interfere with the measurement. Herein, the role of the detection step becomes important where the signal amplification can be achieved via the use of several coatings of antibodies. A color producing substrate, like tetramethylbenzidine (TMB), in staining can be used.

It produces the color based on an enzymatic reaction. The substrate can be chosen based on the instrument available like in case of colorimetric reactions, spectrophotometers are used and in case of a fluorometric tag, fluorimeters are implemented.

14.2.5 Types of ELISA Assay

14.2.5.1 Direct ELISA

This technique allows for coating of Ag directly to wells of microtitre plate which is followed by the addition of enzyme-labeled primary Ab that detects the Ag that is complementary. With this type of assay, an enzyme conjugated antibody is applied for detection or fluorescence moiety is used. After the incubation in presence of the immobilized antibody–antigen complex on the well of the plate, output signal detection is performed when the chromogenic substrate is applied or via exciting and collecting the emission signal of the fluorophore (Fig. 14.4).

Advantages

- Fast and minimal steps involved in the procedure.
- Minimum precursor requirement makes it less error prone.

Disadvantages

- The immobilization of the antigen is not specific due to which background related issues are seen.

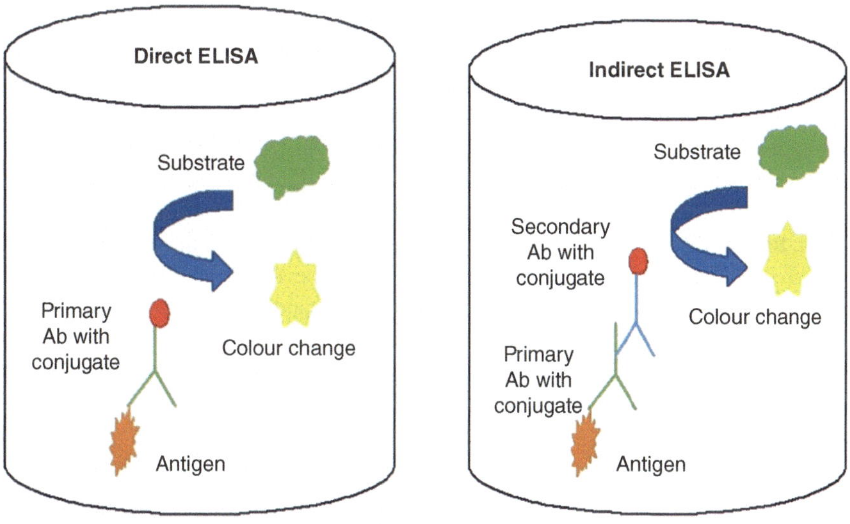

Fig. 14.4 Principle of direct and indirect ELISA

- Offers less flexibility in terms of primary antibody.
- The absence of signal amplification reduces sensitivity.

14.2.5.2 Indirect ELISA

The indirect ELISA method exhibits higher sensitivity since it employs enzyme-labeled secondary antibody which interacts with a primary antibody. The secondary antibody is mostly polyclonal in origin with anti-species reactivity. The indirect ELISA helps in quantifying the total concentration of antibody in samples. In this technique, Ag is coated directly to wells of microtitre plate which is followed by the addition of primary Ab that detects the Ag that is complementary. Then enzyme-labeled secondary antibody is added which will bind to the primary Ab (Fig. 14.4).

Advantages

- Offers high sensitivity and flexibility as the number of secondary antibodies can bind to a primary antibody and one type of secondary antibody can label different primary antibodies.
- It is cheaper as there is a requirement of fewer labeled antibodies.

Disadvantages

- Higher signal to noise ratio.
- Time consuming and extra labor required.

14.2.5.3 Sandwich ELISA

Sandwich ELISA employs the usage of antibody pairs such as capture antibody and detection antibody which can be a monoclonal or polyclonal antibody. Each antibody is highly specific toward epitope of an antigen, and the assay is found to be more suitable for antigens possessing two epitopes. The specificity of matched antibody pairs is very important to confirm their binding to different epitopes in obtaining precise results. As its name indicates, the capture antibody interacts with an antigen which can then be detected via both direct and indirect ELISA techniques. Since the antigen quantification is taking place between upper and lower layer of antibodies, the entire method is termed as sandwich ELISA (Fig. 14.5).

The primary step in sandwich ELISA involves coating of ELISA plate with a capture antibody. To this antibody coated plate, a sample containing antigen is added, followed by antibody detection. The detection antibody can be either enzyme labeled or unlabeled, depending on which it will be direct sandwich ELISA or indirect sandwich ELISA, respectively. In indirect sandwich ELISA, a secondary enzyme-labeled antibody-based detection is introduced that binds to the primary unlabeled detection antibody. Sandwich ELISA is found to be 2–5 times more sensitive when compared to direct and indirect ELISA techniques.

Fig. 14.5 Sandwich ELISA

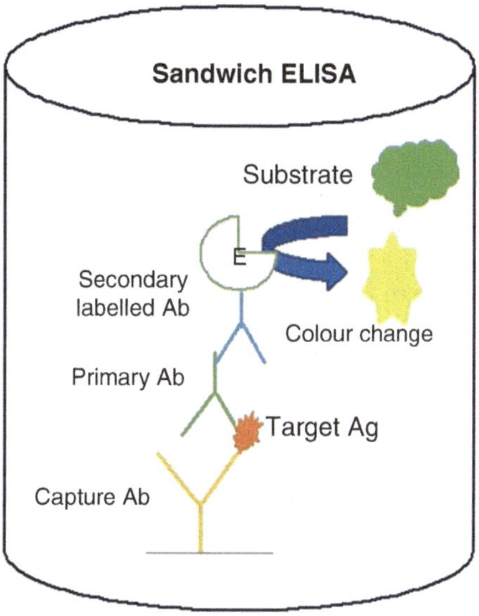

Advantages

- Offers high sensitivity compared to direct or indirect ELISA.
- Introduces highly specific reaction due to the involvement of two antibodies for antigen detection.
- Both direct and indirect technique are implemented in detection.

Disadvantages

- Optimization in terms of antibody becomes problematic due to cross-reactivity issues.
- For recognition of a specific epitope, only monoclonal antibodies can be applied as matched pairs.
- To procure monoclonal antibodies, it is a tedious process in case of matched pairs and is more expensive than polyclonal antibodies.

14.3 Blotting Techniques

Visualization of a specific DNA or RNA or protein fragment out of many molecules requires a technique called blot transfer. In this technique, the separated bands are transferred onto a nitrocellulose membrane from the gel. The following are the types of blotting:

- Southern blotting
- Northern blotting
- Western blotting

Southern blotting is used for the transfer of DNA from gel onto the membrane while Northern and Western blotting are used for the transfer of RNA and protein bands, respectively. One other blotting technique is south-western blotting which examines the protein–DNA interactions.

14.3.1 Southern Blotting

Southern blotting is named after E.M. Southern who discovered this technique in 1975. It is used for the transfer of DNA from gel onto the membrane. The following are the steps involved:

- DNA to be analyzed is digested with a restriction endonuclease.
- Electrophoresis to maximally separate restriction fragments in the expected size range. A set of standards of known size is run in one lane of the gel.
- Blot fragments onto a nitrocellulose membrane.
- Hybridize with the ^{32}P probe.
- Autoradiography.

14.3.1.1 Procedure

In this technique, the sample DNA is digested with restriction enzymes to obtain fragments of different lengths. These differently sized DNA segments are then passed through agarose gel electrophoresis for their separation based on their lengths. The gel so obtained with different bands of DNA fragments is placed on top of buffer saturated filter papers which act as a filter paper wick. Nitrocellulose filter is placed above the gel above which many dry filter paper sheets are placed. With the movement of buffer toward the dry filter papers, the DNA bands are also moved upwards and hence they get bound to the nitrocellulose filter membrane.

Now, the nitrocellulose filter is removed and baked in vacuum. DNA fragments on the nitrocellulose filter are hybridized with single-stranded radioactively labeled probes. Washing is done to remove unbound probes, and finally the DNA bands with radioactivity are visualized by autoradiography (Fig. 14.6).

14.3.1.2 Application

- It is used to identify a specific DNA in a DNA sample.
- Used in the identification of viral infection and certain bacterial infections.
- It is useful in the study of mutation and gene rearrangement, this property is used to diagnose neonatal disease and genetic disease.
- Due to the precision in DNA identification, this technique is used in phylogenetic studies, paternity and maternity analysis, forensic studies, and personal identification.

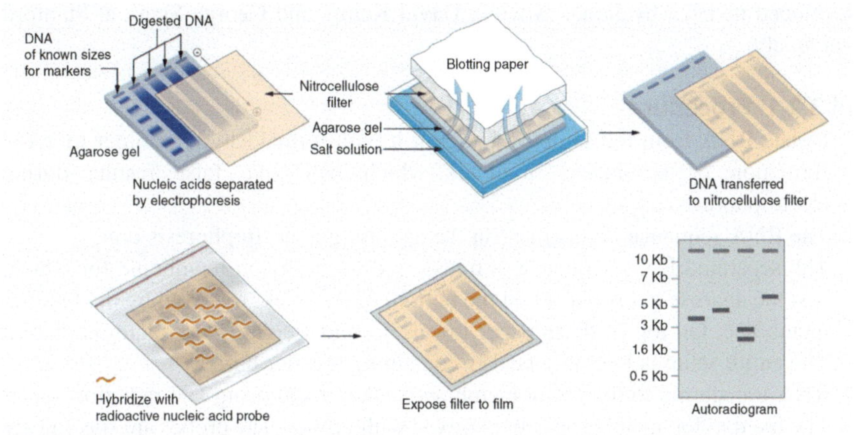

Fig. 14.6 Southern blotting (reproduced with permission from Darryl Leja, National Human Genome Research Institute)

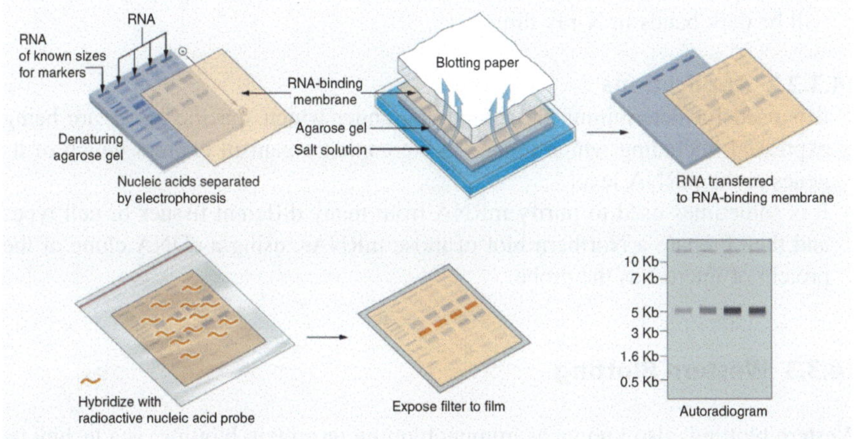

Fig. 14.7 Northern blotting (reproduced with permission from Darryl Leja, National Human Genome Research Institute)

14.3.2 Northern Blotting

In Northern blotting, RNA molecules are blot transferred from the gel onto a chemically reactive paper (Fig. 14.7). It is a technique used for the study of gene expression. It is done by detection of particular RNA (or isolated mRNA). mRNA is generally represented as 5% of the overall RNA sequence. This method reveals the identity, number, activity, and size of the particular gene. This blotting technique can also be used for the growth of a tissue or organism. The Northern blot technique was developed in 1977 by James Alwine, David Kemp, and George Stank at Stanford University.

14.3.2.1 Procedure

1. Isolate RNA from the samples and treat it with formaldehyde in order to avoid formation of secondary structures which will give false results during electrophoresis.
2. The RNA sequence is separated in the agarose gel electrophoresis unit.
3. The separated RNA sequence is transferred to the nylon membrane for which, first the agarose gel is placed on the bottom of the stack, followed by the blotting membrane. On top of these paper towels, a mild weight (glass plate) is placed. The entire setup is kept in a beaker containing transfer buffer.
4. RNA transferred to the nylon membrane is then fixed using UV radiation.
5. The fixed nylon membrane is then mixed with probes. The probes are specifically designed for the gene of interest, so that they will hybridize with RNA sequences on the blot corresponding to the sequence of interest.
6. The blot membrane is washed to remove unwanted probe.
7. Labeled probe is detected by chemiluminescence or autoradiography. The result will be dark bands in X-ray film.

14.3.2.2 Applications

- It is used for determining the conditions under which specific genes are being expressed, including which tissues in a complex organism express which of its genes at the mRNA level.
- It is sometimes used to purify mRNA from many different tissues or cell types and then prepare a Northern blot of those mRNAs, using a cDNA clone of the protein of interest as the probe.

14.3.3 Western Blotting

Western blotting, also known as immunoblotting or protein blotting, is a technique used to detect the presence of a specific protein in a complex protein mixture. It is used for separation or identification of protein molecules (Fig. 14.8). This technique can be used for both the active 3D protein and denatured long peptide chains. The 3-D protein in its active structure has sulfur-hydroxyl bonds in the structure. This methodology classifies protein based on the molecular weight and charge.

14.3.3.1 Procedure

- Extraction of cellular proteins from a complex mixture of intracellular and extracellular proteins (from tissue, cells, etc.)
- Quantification of protein concentration and electrophoretic separation of proteins within a gel matrix.
- Transfer of gel to a membrane with a high affinity for proteins like nitrocellulose membrane and polyvinylidene fluoride (PVDF) membrane.

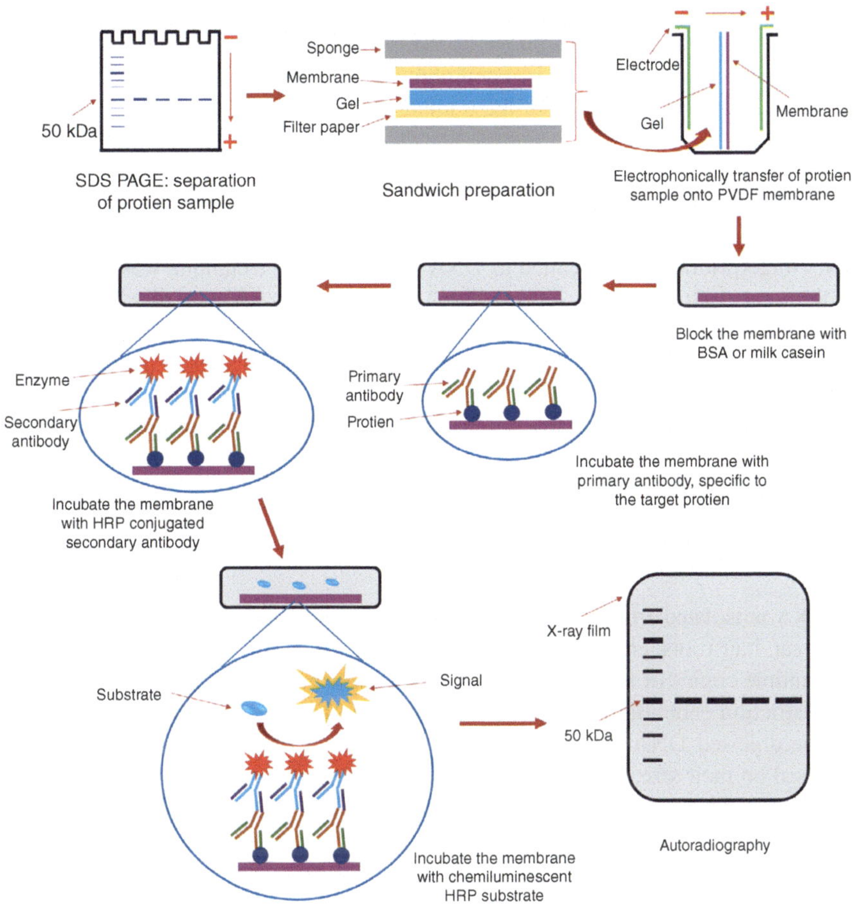

Fig. 14.8 Western blotting (reproduced from Roy et al. 2019 with permission from Elsevier)

- "Blocking" the membrane to reduce non-specific binding of Ab by using blocking solutions containing 3–5% BSA or non-fat dried milk in a solution of PBS (phosphate buffered saline) or TBS (tris buffered saline)
- Ab probing is done by adding Ab to the membrane by incubating the membrane with primary Ab.
- Excess Abs are removed by washing.
- Membrane is incubated with secondary Ab linked to a label such as horseradish peroxidase.
- Secondary Ab will bind to primary Ab bound with the protein of the membrane.
- Development and detection of the signal, which is theoretically proportional to the degree of antigen/antibody binding.
- Quantification of the resulting bands.

14.3.3.2 Applications

- Detection of viral proteins.
- Characterization of monoclonal and polyclonal antibody preparations.
- Determining the specificity of the immune response to viral antigens.

14.4 DNA Fingerprinting

DNA fingerprinting is also called as DNA profiling. This technique was discovered in the year 1986 by Alec Jeffrey. DNA fingerprinting aids in identification of individuals at the genetic level by characterizing the segments of DNA. It is a well-known fact that every living organism can be differentiated from the other only due to the sequence of nucleotides in the chromosome. In India, facilities for DNA fingerprinting are available at Centre for DNA Fingerprinting and Diagnostics, Hyderabad.

14.4.1 Procedure

- DNA is isolated from samples such as blood stains, semen, urine, tears, saliva, sweat, hair roots, etc.
- Multiple copies of isolated DNA are produced using PCR.
- Restriction endonucleases are used to cut DNA at specific sites by digestion.
- The digested DNA is subjected to electrophoresis in which DNA is separated based on their size.
- Now southern blotting is performed from which an X-ray film is produced with DNA in different bands.
- Bands produced by DNA of different samples are compared.

The process of DNA fingerprinting is presented in Figs. 14.9 and 14.10. Mini-satellites which are tandem repeats of short sequences are used as probes while preparing DNA fingerprints of humans. In case of plants, such probes are not present, therefore, usually RFLPs (Restriction Fragment Length Polymorphism) or simple sequence repeats like (CT) or (AC), etc. are used as probes for DNA fingerprinting for varietal identification.

This technique of DNA fingerprinting has revolutionized the field of forensic medicine as it is very beneficial for identification of criminals like murderers or rapists. This also helps in solving parentage disputes by identifying the real biological father of the child by analyzing the DNA fingerprints of child and suspected father (Fig. 14.11).

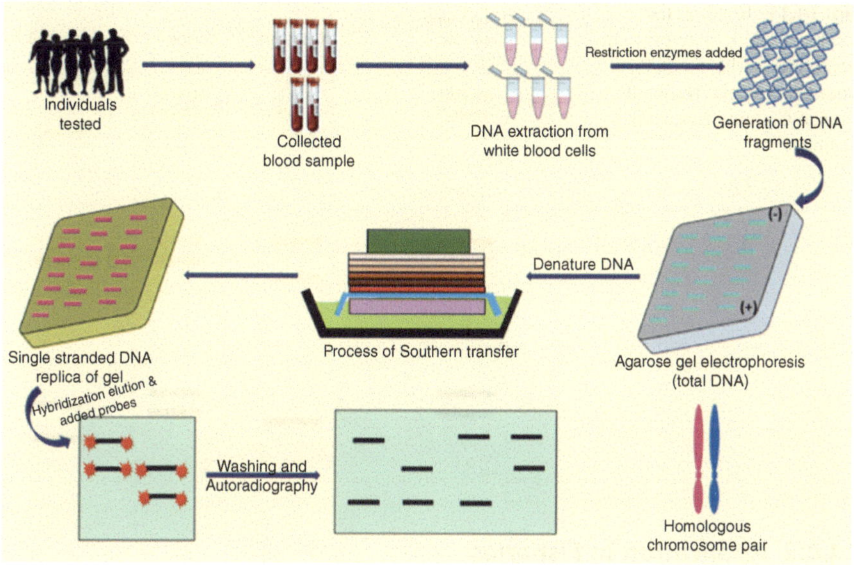

Fig. 14.9 Process of DNA fingerprinting (reproduced from Sinha et al. 2018 with permission from Springer Nature)

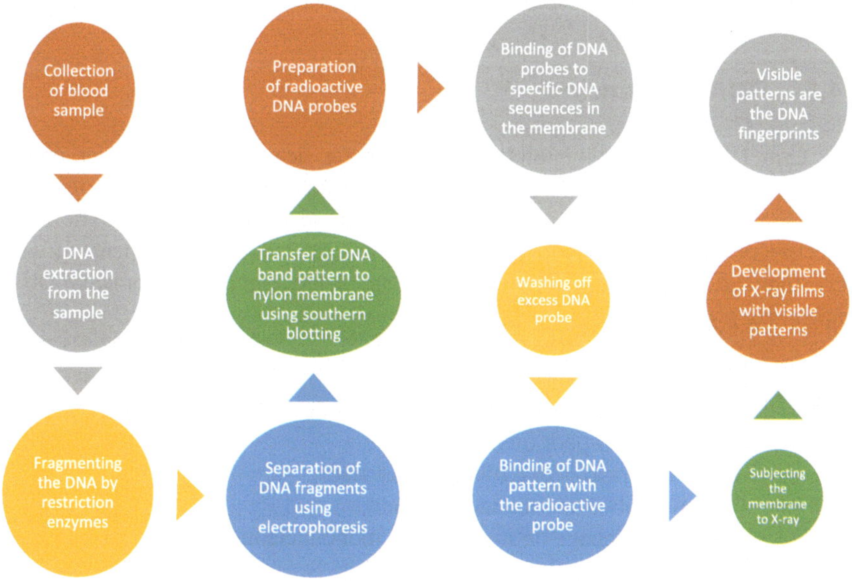

Fig. 14.10 Detailed steps involved in DNA fingerprinting

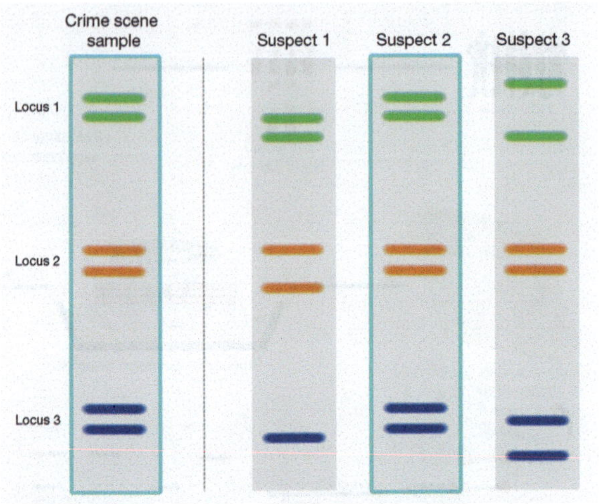

Fig. 14.11 Reading the results of DNA fingerprinting (reproduced with permission from Darryl Leja, National Human Genome Research Institute)

14.4.2 Application in Fisheries

- Used to verify the identity of cultured cell lines and various lines of clonal fishes.
- Parents can be identified which helps to avoid inbreeding.
- Genetic improvement using gene linkage analysis is possible.

14.5 DNA Sequencing

The term DNA sequencing refers to methods for determining the order of the nucleotides bases adenine, guanine, cytosine, and thymine in a molecule of DNA. DNA sequencing is very significant in research and forensic science. The main objective of DNA sequence generation method is to evaluate the sequencing with very high accuracy and reliability. There are some common automated DNA sequencing problems:

- Failure of the DNA sequence reaction.
- Mixed signal in the trace (multiple peaks).
- Short read lengths and poor quality data.
- Excessive free dye peaks "dye blobs" in the trace.
- Primer dimer formation in sequence reaction.
- DNA polymerase slippage on the template mononucleotide regions.

A sequencing can be done by different methods:

- Maxam–Gilbert sequencing.
- Chain termination method/Sanger method.
- Dye-terminator sequencing.

- Pyrosequencing.
- High-throughput sequencing.

14.5.1 Maxam–Gilbert Sequencing

In the late 1970s, Allan Maxam and Walter Gilbert developed this sequencing method based on chemical degradation. In this method, the DNA fragment to be sequenced is radioactively labeled at one end which is subjected to random cleavage using chemicals at any of the four base positions (A,T,G,C). The chemical agent cleaves based on three steps such as base modification, removal of the modified base from its sugar, and DNA strand breaking at that sugar position (Maxam and Gilbert 1977). The resultant radiolabeled products are separated using polyacrylamide gel electrophoresis and are visualized using auto radiography. However, the chemical reactions are slow, and the use of hazardous chemicals requires special handling care. Some of the conditions for which this method can be used is given below:

- For sequencing the genome from which details on DNA methylation and chromatin structure can be obtained (Church and Gilbert 1984)
- The accuracy of synthesized oligonucleotides can be confirmed (Ornstein and Kashdan 1985)
- DNA regions with hairpin loops can be verified (Ornstein and Kashdan 1985)
- Rare bases, such as Hoogsteen base pairs can be located (Sayers and Waring 1993)
- Point mutations can be detected (Ferraboli et al. 1993);
- Ambiguities arising during dideoxy sequencing can be solved (Goszczynski and McGhee 1991);
- DNA–protein interactions can be analyzed (Isola et al. 1999);

14.5.2 Chain Termination Method

It was developed by Frederick Sanger in the year 1977 at Cambridge University. Hence it is also called Sanger sequencing. The key principle of the Sanger method was the use of dideoxynucleotide triphosphates (ddNTPs) as DNA chain terminators. Therefore it is also called as dideoxy sequencing. In this method, a primer binds to a denatured DNA molecule and initiates the synthesis of a single-stranded polynucleotide in the presence of a DNA polymerase enzyme. Thus, a covalent bond forms between the 3′ carbon atom of the deoxyribose sugar molecule in one nucleotide and the 5′ carbon atom of the next.

14.5.2.1 Requirements

- Single-stranded DNA template
- DNA primer
- DNA polymerase

- Radioactively or fluorescently labeled nucleotides (dNTPs)
- Modified nucleotides (ddNTPs)

14.5.2.2 Procedure

The DNA sample is divided into four separate sequencing reactions, containing all four standard deoxynucleotides (dATP, dGTP, dCTP, dTTP) and the DNA polymerase. To each reaction is added only one of the four dideoxynucleotide (ddATP, ddGTP, ddCTP, ddTTP) which are the chain terminating nucleotides, lacking a 3′-OH group required for the formation of a phosphodiester bond between two nucleotides, thus terminating DNA strand extension and resulting in DNA fragments of varying length. The newly synthesized and labeled DNA fragments are heat denatured and separated according to their size by gel electrophoresis with each of the four reactions run in one of the four individual lanes (lanes A, T, G, C). The DNA bands are then visualized by autoradiography or UV light, and the DNA sequence can be directly read off the X-ray film or gel image (Fig. 14.12). A dark band in a lane indicates a DNA fragment that is result of chain termination after incorporation of a ddNTP. The relative position of the different bands among the four lanes is then used to read (from bottom to top) the DNA sequence.

The technical variations of chain termination sequencing include tagging with nucleotides containing radioactive phosphorus for labeling, or using a primer labeled at the 5′ end with a fluorescent dye. Dye- primer sequencing facilitates reading in an optical system for faster and more economical analysis and automation. Chain termination methods have greatly simplified DNA sequencing. Limitations include non-specific binding of the primer to the DNA, affecting accurate read-out of the DNA sequence and DNA secondary structures affecting the fidelity of the sequence.

14.5.3 Dye-Terminator Sequencing

Dye-terminator sequencing utilizes labeling of the chain terminator ddNTPs, which permits sequencing in a single reaction rather than four reactions as in the labeled primer method. In this, each of the four dideoxy nucleotide chain terminators is labeled with fluorescent dyes each with different wavelengths of fluorescence and emission.

14.5.4 Pyrosequencing

This method is based on real-time detection of PPi released during DNA polymerization reaction (Franca et al. 2002). Initially, each dNTP is tried in the nucleic acid polymerization reaction. PPi is released when one of the deoxynucleotides (dATP, dCTP, dGTP, or dTTP) is incorporated into the chain extension by DNA polymerase.

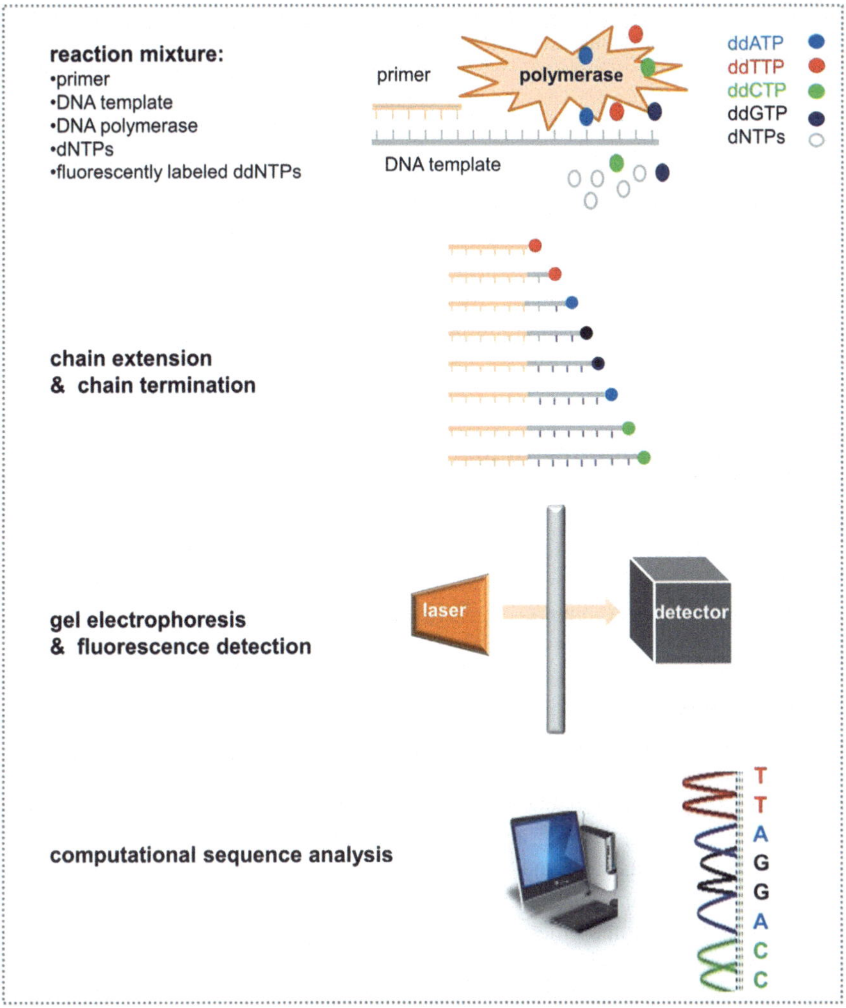

Fig. 14.12 Sanger sequencing (reproduced from Cheng et al. 2015 with permission from Elsevier)

$$(DNA)_n + dNTP \xrightarrow{\text{DNA polymerase}} (DNA)_{n+1} + PPi$$

The PPi produced is converted into ATP with the help of adenosine 5′--phosphosulfate (APS) in the presence of ATP sulphurylase.

$$PPi + APS \xrightarrow{\text{ATP sulphurylase}} ATP + SO_4^{-2}$$

The enzyme luciferase catalyzes luciferin into oxyluciferin and emits a photon (*hv*) by the bioluminescent reaction.

$$ATP+luciferin+O_2 \xrightarrow{luciferase} AMP+PPi+oxyluciferin+CO_2+hv$$

Depending on the number of dNTPs added per chain, the number of photons emitted per template chain will vary. The sequence can then be determined by simply noting if incorporations occur and by counting the number of incorporations (by measuring the light intensity). The amount of light emitted can be measured by an avalanche photodiode, photomultiplier tube, or with a charged-coupled device camera (Franca et al. 2002) (Fig. 14.13).

14.5.5 High-Throughput Sequencing

High-throughput sequencing includes modern sequencing technologies such as Illumina, Ion Torrent, Pacific Biosciences, and Oxford Nanopore Technologies. They are also referred as next-generation, third-generation, and fourth-generation sequencing methods (Pradhan et al. 2019) and cyclic array sequencing (Shendure et al. 2008). They are fast, cost effective and generate massive amount of data.

14.5.5.1 Illumina

Illumina/Solexa released the Genome Analyzer II in 2006 and presently they are the dominant tool used for high-throughput sequencing (Reuter et al. 2015). The overall error rates in Illumina sequencing are below 1%, and substitution is the most common type of error (Dohm et al. 2008). In Illumina sequencing, the adaptor-ligated DNA fragments are clonally amplified on the surface of a glass slide (Bentley et al. 2008). The amplified bases are read using a cyclic reversible termination strategy, which sequences the template strand one nucleotide at a time through progressive rounds of base incorporation, washing, imaging, and cleavage. The polymerization reaction is paused using fluorescently labeled 3′-O-azidomethyl dNTPs which enables removal of unincorporated bases and the fluorescent imaging is used to determine the added nucleotide (Guo et al. 2008). Following scanning of the flow cell with a coupled-charge device (CCD) camera, the fluorescent moiety and the 3′ block are removed, and the process is repeated (Fig. 14.14).

14.5.5.2 Ion Torrent

Life Technologies commercialized Ion Torrent's semiconductor sequencing technology benchtop Ion PGM sequencer in 2010 followed by Ion Proton in 2012 (Reuter et al. 2015). The PGM is most useful for targeted resequencing projects and small genome analysis, whereas the proton is capable of exome sequencing and whole-transcriptome analysis. Insertions and deletions (indels) are the most common error types (Liu et al. 2012). The adapter-ligated DNA fragments are clonally amplified on the surface of beads using emulsion-PCR. Then the beads are

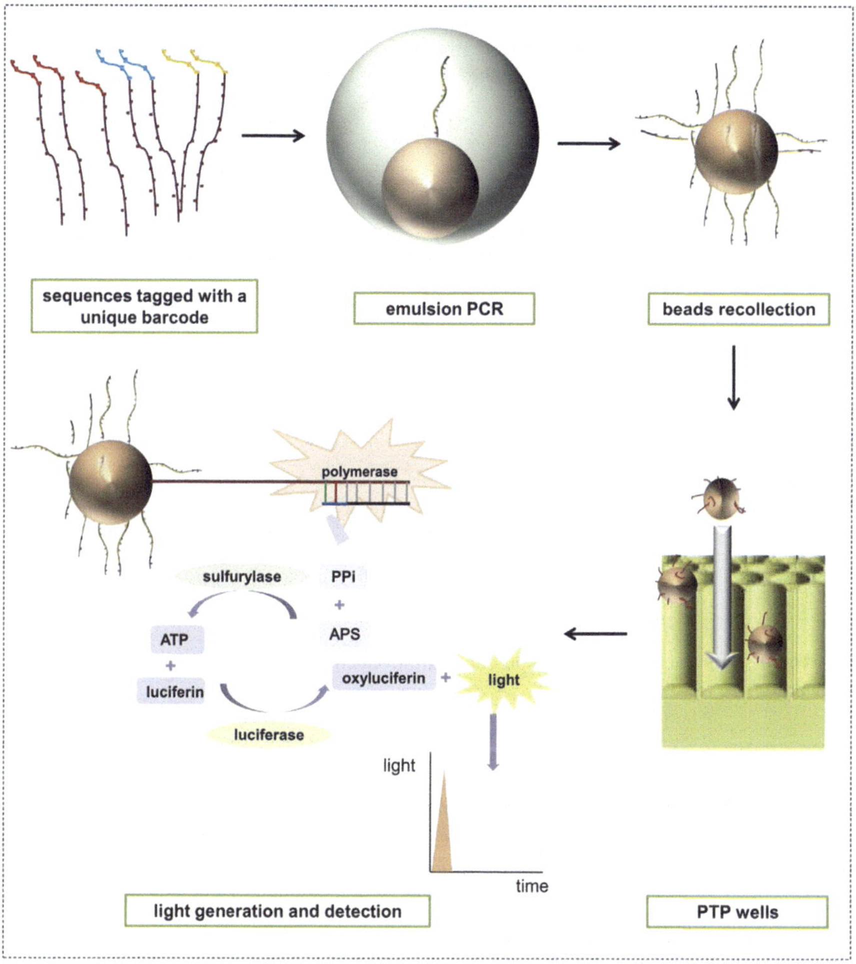

Fig. 14.13 Pyrosequencing (reproduced from Cheng et al. 2015 with permission from Elsevier)

distributed into microwells where a sequencing-by-synthesis reaction occurs. The semiconductor sequencing measures pH changes induced by the release of hydrogen ions during DNA extension (Rothberg et al. 2011). The bottom of the microwell has a sensor which detects the change in pH and converts it into a voltage signal. The voltage signal is proportional to the number of bases incorporated, and the sequential addition of individual nucleotides during each sequencing cycle allows base discrimination.

14.5.5.3 Pacific Biosciences

Pacific Biosciences commercialized single-molecule real-time (SMRT) sequencing which was pioneered by Nanofluidics, Inc. (Reuter et al. 2015). In this method,

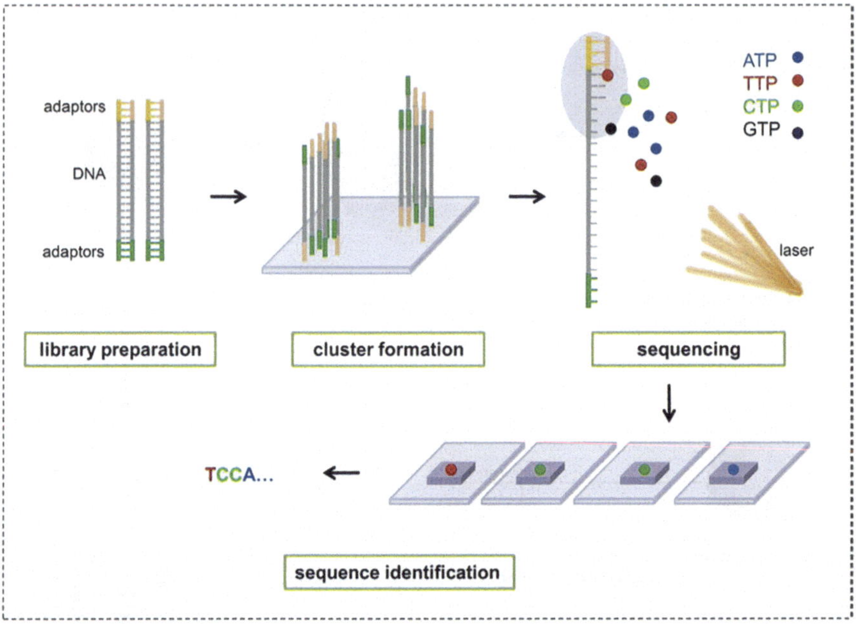

Fig. 14.14 Illumina sequencing (reproduced from Cheng et al. 2015 with permission from Elsevier)

template is prepared by ligating the single-stranded, hairpin adapters onto the ends of digested DNA or cDNA molecules which generates a capped template (SMRT-bell). By using a strand displacing polymerase, the original DNA molecule can be sequenced multiple times, thereby increasing accuracy (Travers et al. 2010). SMRT sequencing allows direct sequencing of native, and potentially modified, DNA and avoids clonal amplification. DNA synthesis occurs in zeptoliter-sized chambers, called zero-mode waveguides (ZMW), in which a single polymerase is immobilized at the bottom of the chamber. This chamber reduces background noise so that phosphate-labeled versions of all four nucleotides can be present simultaneously. Thus, polymerization occurs continuously, and the DNA sequence can be read in real time from the fluorescent signals recorded in a video (Eid et al. 2009). Indels are the common errors which amount to high error rates of ~11% for single pass reads.

14.5.5.4 Oxford Nanopore Technologies

Oxford nanopore technologies developed and commercialized the nanopore-based sequencing which is another single-molecule sequencing method and hence the error rates are higher. MinION, a USB-powered, portable sequencer was released in 2014 for nanopore sequencing (Reuter et al. 2015). Nanopore sequencing relies on the transition of DNA or individual nucleotides through a small channel (Wang et al. 2015). The technology consists of sequencing flow cell which has number of independent microwells, each containing a synthetic bilayer perforated by biologic nanopores. When the bases are threaded through the pore by a molecular motor protein, the bases induce changes in current which are measured and thereby the sequencing is accomplished. Library is prepared by fragmentation of DNA and ligation of adapters which can be done with or without PCR amplification. The first adapter is bound with a propriety motor enzyme as well as a molecular tether, whereas the second adapter is a hairpin oligonucleotide that is bound by a second HP motor protein (Quick et al. 2014). This allows sequencing of both strands of DNA from a single molecule, which increases accuracy (Quick et al. 2014).

14.5.6 Applications of DNA Sequencing Technologies (Mardis 2017)

- The genetic information of a particular DNA segment or a whole genome can be understood.
- The genes and regulatory instructions that are contained in the DNA molecule can be determined.
- The DNA sequence can be screened for characteristic features of genes, such as open reading frames (ORFs) and CpG islands.
- Homologous DNA sequences from different organisms can be compared for evolutionary analysis between species or populations.
- DNA sequencing can reveal changes in a gene that may cause a disease.

14.6 Protein Sequencing

Proteins are amino acids which are arranged in specific sequences. The amino acid sequences are determined by the sequence of bases in mRNA. Sanger was the first to sequence the protein from insulin, and he established that each protein has a unique sequence which determines its structure and functions. Some proteins are made up of more than one polypeptide chain, and the order of amino acids in a polypeptide chain can be analyzed by sequencing the proteins. Protein sequencing can be carried out by two direct techniques such as Edman degradation and mass spectrometry.

14.6.1 Protein Sequencing by Edman Degradation

This method was developed by Pher Edman in 1960. The proteins are first separated on SDS-PAGE gel and blotted on to a solid support like PVDF membrane. The blotted PVDF membrane is introduced into the automatic protein sequencer (Saraswathy and Ramalingam 2011). When the proteins on the solid support are exposed to the phenyl isothiocyanate (Edman reagent), the sample is incubated with an anhydrous acid (e.g., trifluoroacetic acid) that breaks the peptide bond between the first and second amino acids of the protein. Thus, the amino acids at the N-termini residue become free as a thiazolinone derivative. This thiazolinone is then extracted using an organic solvent, and dried. Finally, it is converted to a more stable phenylthiohydantoin (PTH) derivative. The PTH derivatives are separated using HPLC and are determined by their order and elution location (Büyükköroğlu et al. 2018).

14.6.1.1 Advantages of Edman Degradation

- Requires small quantity (in pg) of sample.
- Highly sensitive and reliable.
- Direct analysis of intact proteins without digestion.

14.6.1.2 Disadvantages of Edman Degradation (Saraswathy and Ramalingam 2011)

- Proteins stained with silver nitrate cannot be sequenced.
- This method can detect only 10 amino acids in 24 h.
- In a single reaction, only 50 amino acids can be sequenced.

14.6.2 Mass Spectrometry (MS)

It is a high-throughput proteomic technique and is a de novo sequencing method which determines the amino acid sequence from the mass spectrum (Saraswathy and Ramalingam 2011). The basic principle involved is that the molecules are separated based on their mass-to-charge (m/z) ratio by ionization of the molecules with high energy electrons to break a molecule into fragments. Fragmentation pattern and m/z values provide information about the molecular weights and chemical structures of the peptides and proteins (Büyükköroğlu et al. 2018). The measurements are done on ionized molecules in the gaseous phase and hence molecules from liquid or solid phase must be transferred into gaseous phase (Ong and Mann 2005).

According to Oudenhove and Devreese (2013), a spectrometer consists of a device for sample input into the machine, molecular ionization source, mass analyzer, detector, vacuum system, and a computer-based data obtaining and processing system. Initially, the samples are introduced into the spectrometer and are subjected to molecular ionization for conversion into gaseous phase. This molecular ionization is carried out by two methods such as direct insertion with a probe or plate with matrix-assisted laser desorption/ionization (MALDI)-MS and direct infusion or injection into the ionization source with electrospray ionization (ESI)-MS (Kang 2012). However, MALDI-MS is a widely used method (Lagarrigue et al. 2012; Gessel et al. 2014).

After ionization, the ionized molecules are introduced into the mass analyzer for separation of molecules based on their m/z ratio by electric and/or magnetic fields or by measuring the time an ion takes to reach a fixed distance from the point of ionization to the detector (Rompp and Spengler 2013). The separated ions from mass analyzer is monitored and recorded by the detector. After detection of ions, the signals are recorded on a graph by plotting the amount of signal versus m/z ratio which shows the presence and abundance of different molecular size peptides and proteins and the energy level of the molecules of a particular kind. MS vacuum pumps are used for free movement of ions within the spectrometer, and computer-based data obtaining and processing system is used for obtaining and processing data (Büyükköroğlu et al. 2018).

14.6.3 Applications of Protein Sequencing

- Analysis of structure and function of protein.
- Identification of new protein biomarkers and drugs.
- Construction of phylogenetic tree to find the evolutionary relationship.
- Orthologous and paralogous proteins can be identified.
- Probe sequence for cDNA library screening.
- Protein engineering studies.

References

Bentley DR, Balasubramanian S, Swerdlow HP, Smith GP, Milton J, Brown CG, Hall KP, Evers DJ, Barnes CL, Bignell HR et al (2008) Accurate whole human genome sequencing using reversible terminator chemistry. Nature 456:53–59

Büyükköroğlu G, Dora DD, Özdemir F, Hızel C (2018) Techniques for protein analysis. Omics technologies and bio-engineering, pp 317–351. https://doi.org/10.1016/b978-0-12-804659-3.00

Cheng L, Cheng X, Deng M et al (2015) Techniques for oral microbiology. In: Atlas of oral microbiology. Elsevier, pp 15–40. https://doi.org/10.1016/b978-0-12-802234-4.00002-1

Church GM, Gilbert W (1984) Genomic sequencing. Proc Natn Acad Sci USA 81:1991–1995

Dohm JC, Lottaz C, Borodina T, Himmelbauer H (2008) Substantial biases in ultra-short read data sets from high-throughput DNA sequencing. Nucleic Acids Res 36:e105

Eid J, Fehr A, Gray J, Luong K, Lyle J, Otto G, Peluso P, Rank D, Baybayan P, Bettman B et al (2009) Real-time DNA sequencing from single polymerase molecules. Science 323:133–138

Ferraboli S, Negri R, Di Mauro E, Barlati S (1993) One-lane chemical sequencing of 3′ fluorescent-labeled DNA. Anal Biochem 214:566–570

Franca LTC, Carrilho E, Kist TBL (2002) A review of DNA sequencing techniques. Q Rev Biophys 35(2):169–200. https://doi.org/10.1017/S0033583502003797

Gessel MM, Norris JL, Caprioli MR (2014) MALDI imaging mass spectrometry: spatial molecular analysis to enable a new age of discovery. J Proteome 107:71–82

Goszczynski B, McGhee JD (1991) Resolution of sequencing ambiguities: a universal FokI adapter permits Maxam-Gilbert re-sequencing of single stranded phagemid DNA. Gene 104:71–74

Guo J, Xu N, Li Z, Zhang S, Wu J, Kim DH, Sano Marma M, Meng Q, Cao H, Li X et al (2008) Four-color DNA sequencing with 3'-O-modified nucleotide reversible terminators and chemically cleavable fluorescent dideoxynucleotides. Proc Natl Acad Sci U S A 105:9145–9150

Isola NR, Allman SL, Golovlov VV, Chen CH (1999) Chemical cleavage sequencing of DNA using matrix-assisted laser desorption/ionization time-of-flight mass spectrometry. Anal Chem 71:2266–2269

Kang JS (2012) Principles and applications of LC-MS/MS for the quantitative bioanalysis of analytes in various biological samples. In: Prasain JK (ed) Tandem mass spectrometry—applications and principles. The Author(s), pp 441–492

Lagarrigue M, Lavigne R, Gue'vel B, Com E, Chaurand P, Pineau C (2012) Matrix-assisted laser desorption/ionization imaging mass spectrometry: a promising technique for reproductive research. Biol Reprod 86(3):74. 1–11

Liu S, Zhang Y, Sun F, Jiang Y, Wang R, Li C, Zhang J, Liu Z (2012) Functional genomics research in aquaculture: principles and general approaches. In: Saroglia M, Liu Z(J) (eds) Functional genomics in aquaculture. John Wiley & Sons Inc, pp 1–40

Mardis ER (2017) DNA sequencing technologies: 2006–2016. Nat Protoc 12(2):213

Maxam AM, Gilbert W (1977) A new method for sequencing DNA. Proc Natn Acad Sci USA 74: 560–564

Ong SE, Mann M (2005) Mass spectrometry-based proteomics turns quantitative. Nat Chem Biol 1: 252–262

Ornstein DL, Kashdan MA (1985) Sequencing DNA using ^{35}S-labeling: a troubleshooting guide. BioTechniques 3:476–484

Oudenhove LV, Devreese B (2013) A review on recent developments in mass spectrometry instrumentation and quantitative tools advancing bacterial proteomics. Appl Microbiol Biotechnol 97:4749–4762

Pradhan D, Kumar A, Singh H, Agrawal U (2019) Chapter 4 – high-throughput sequencing. In: Misra G (ed) Data processing handbook for complex biological data sources. Academic Press, pp 39–52., ISBN 9780128165485. https://doi.org/10.1016/B978-0-12-816548-5.00004-6

Quick J, Quinlan AR, Loman NJ (2014) A reference bacterial genome dataset generated on the MinION™ portable single-molecule nanopore sequencer. GigaScience 3:1–6. Open Access

Reuter JA, Spacek DV, Snyder MP (2015) High-throughput sequencing technologies. Mol Cell 58(4):586–597. https://doi.org/10.1016/j.molcel.2015.05.004

Rompp A, Spengler B (2013) Mass spectrometry imaging with high resolution in mass and space. Histochem Cell Biol 139:759–783

Rothberg JM, Hinz W, Rearick TM, Schultz J, Mileski W, Davey M, Leamon JH, Johnson K, Milgrew MJ, Edwards M et al (2011) An integrated semiconductor device enabling non-optical genome sequencing. Nature 475:348–352

Roy J, Jain N, Singh G, Das B, Mallick B (2019) Small RNA proteome as disease biomarker: an incognito treasure of clinical utility. In: AGO-driven non-coding RNAs, pp 101–136. https://doi.org/10.1016/B978-0-12-815669-8.00005-1

Saraswathy N, Ramalingam P (2011) Protein sequencing techniques. In: Concepts and techniques in genomics and proteomics, pp 193–201. https://doi.org/10.1533/9781908818058.193

Sayers EW, Waring MJ (1993) Footprinting titration studies on the binding of echinomycin to DNA incapable of forming Hoogsteen base-pairs. Biochemistry 32:9094–9107

Shendure JA, Porreca GJ, Church GM (2008) Overview of DNA sequencing strategies. Curr Protoc Mol Biol 7(81):Unit 7.1. https://doi.org/10.1002/0471142727.mb0701s81

Sinha M, Arjun Rao I, Mitra M (2018) Molecular basis of identification through DNA fingerprinting in humans. In: Dash H, Shrivastava P, Mohapatra B, Das S (eds) DNA fingerprinting: advancements and future endeavors. Springer, Singapore. https://doi.org/10.1007/978-981-13-1583-1_7

Travers KJ, Chin CS, Rank DR, Eid JS, Turner SW (2010) A flexible and efficient template format for circular consensus sequencing and SNP detection. Nucleic Acids Res 38:1–8

Wang Y, Yang Q, Wang Z (2015) The evolution of nanopore sequencing. Front Genet 5:1–20

15 Molecular Markers

A molecular marker is defined as a particular segment of DNA that is representative of the differences at the genome analysis. Molecular markers are also called as genetic markers. In general, molecular markers are sites where differences in DNA sequences occur among members of the same species and they reveal polymorphisms at the DNA level. These markers can be in either coding or non-coding regions.

15.1 Advantages of Molecular Markers

- The markers are stable and detectable in all tissues regardless of growth, differentiation, development, or defence status of the cell.
- They are not influenced by the environment, pleiotropic, and epistatic effects.
- They are used to examine the relationships among individuals, populations, or phylogenetic taxa.
- They are used in screening programs for selecting disease resistant clones and introgression of several resistant genes into a particular background.

15.2 Properties of Ideal Genetic Markers

A good genetic marker must have the following characteristics. They must be

- Co-dominant in expression.
- Single copy.
- Highly polymorphic.
- Easily assayed.
- Highly reproducible.
- Frequently occur in genome.

C. J. Betsy, C. Siva, *Fisheries Biotechnology and Bioinformatics*,
https://doi.org/10.1007/978-981-99-6991-3_15

- Multi-functional.
- Genome-specific in nature.
- Multiplexed.
- Automated.
- Easily available.
- Economical.
- It should not have any detrimental effect on phenotype.

15.3 Types of Markers

Generally, markers are classified as follows:

- Non-PCR-based/hybridization-based markers.
- PCR-based/amplification-based markers.
- Sequence-based markers.

15.3.1 Non-PCR-Based Markers

They are also called as hybridization-based markers. DNA profiles are visualized by hybridizing the restriction enzymes digested DNA to a labeled probe, which can be DNA fragment of known origin or sequence. Restriction Fragment Length Polymorphism comes under this category.

15.3.1.1 Restriction Fragment Length Polymorphism (RFLP)

RFLP is a technique in which organisms may be differentiated by analysis of patterns derived from cleavage of their DNA. It is a technique that exploits variations in homologous DNA sequences. RFLP is defined by the existence of alternative alleles associated with restriction fragments that differ in size from each other. Simply, the variations in the restriction DNA fragments length between individuals of a species are called RFLP.

The basic technique of identifying such RFLP involves fragmenting a sample of DNA by a restriction enzyme, which can recognize and cut DNA wherever a specific short sequence occurs, in a process known as a restriction digest. The resulting DNA fragments are then separated by the length through a process known as agarose gel electrophoresis and transferred to a membrane via the Southern blot procedure. Hybridization of the membrane to a labeled DNA probe then determines the length of the fragments which are complementary to the probe. An RFLP occurs when the length of a detected fragment varies between individuals. Although now largely obsolete due to the rise of inexpensive DNA sequencing technologies, RFLP analysis was the first DNA profiling technique inexpensive enough to seek widespread application.

Principle of RFLP

If two organisms differ in the distance between sites of cleavage of a particular restriction endonuclease, the length of the fragments produced will differ when the DNA is digested with a restriction enzyme. The similarity and differences of the patterns thus generated can be used to differentiate species (and even strains) from one another.

Steps Involved in RFLP

1. The first step in this process is to isolate the DNA from the target.
2. Once the DNA is isolated from the sample, it is subjected to restriction digestion using restriction enzymes.
3. The digested DNA sample is then subjected to gel electrophoresis, in which the DNA is separated based on its size. Many DNA fragments with slight differences in length are produced.
4. The gel is then exposed to a chemical to denature double-stranded DNA to become single- stranded.
5. This is followed by southern blotting where DNA is transferred from gel to nylon membrane.
6. The nylon membrane is then exposed to solution with radioactive complementary nucleotide probes that hybridize to specifically chosen DNA sequences on nylon membrane.
7. The membrane is then placed against X-ray film, where hybridized radioactive probes cause exposure of X-ray film, producing an autoradiogram (Fig. 15.1).
8. RFLP analysis is carried out to detect differences in pattern to confirm polymorphisms.

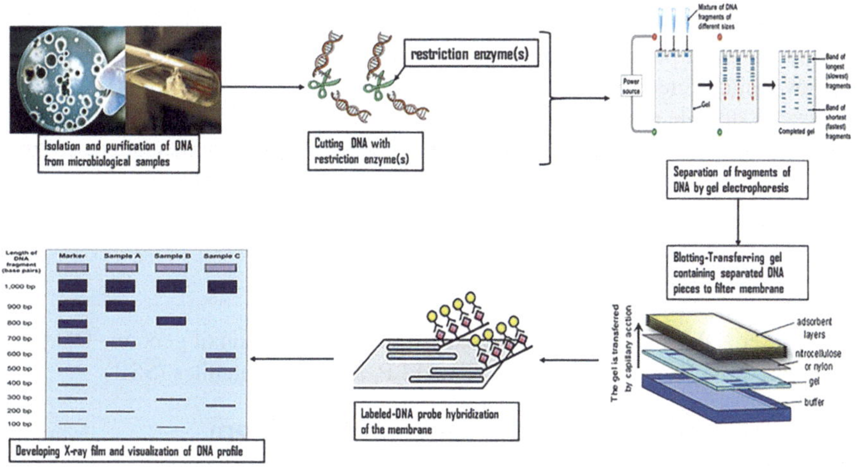

Fig. 15.1 Steps involved in RFLP (reproduced from Muwonge et al. 2019 with permission from Springer Nature)

Applications of RFLP

RFLP analysis was formerly an important tool in genome mapping, localization of genes for genetic disorders, determination of risk for disease, and paternity testing. RFLP can be used in many different settings to accomplish different objectives:

1. In paternity cases or criminal cases to determine the source of a DNA sample. (i.e., it has forensic applications).
2. Determining the disease status of an individual. (e.g., it can be used in the detection of mutations).
3. To measure recombination rates which can lead to a genetic map with the distance between RFLP loci.
4. In the characterization of genetic diversity or breeding patterns in animal populations.
5. RFLP has been developed for chromosomes mapping of humans, mice, maize, tomato, rice, etc.

Advantages of RFLP

- The main advantage of RFLP analysis over PCR-based protocols is that no prior sequence information, nor oligonucleotide synthesis, is required.
- Results are based on reliable genotypic characteristics' rather than on phenotypes.
- It is the co-dominant marker, thus can estimate heterozygosity.
- It is very useful for studying genomic DNA Sequence.
- Highly robust methodology with good transferability between laboratories.

Limitations of RFLP

- Slow.
- Cumbersome.
- Requires a large amount of sample DNA.
- Automation not possible.
- Low levels of polymorphism in some species.
- Few loci detected per assay.
- Need a suitable probe library.
- Needing the combined process of probe labeling, DNA fragmentation, electrophoresis, blotting, hybridization, washing, and autoradiography.

15.3.2 PCR-Based Markers

The markers which are developed based on PCR amplification of genomic DNA are PCR-based markers. It includes RAPD, AFLP, and microsatellites (SSR).

15.3.2.1 Random Amplified Polymorphic DNA (RAPD)

The RAPD technique is based on PCR amplification of genomic DNA with either a single or multiple short oligonucleotide primers of an arbitrary or random sequence (Williams et al. 1990; Krawczyk and Kur 2018). The analysis utilizes short PCR primers consisting of random sequences usually in the size range of 8–15 nucleotides in length (Butler 2012). These oligonucleotides serve as both forward and reverse primer which can amplify fragments from 1 to 10 genomic sites simultaneously. The resulting amplified fragments are in the size range of 0.5–5 kb which are separated by agarose gel electrophoresis. Then it is stained using ethidium bromide, and the polymorphisms are detected as the presence or absence of bands of particular sizes. Variation in the primer annealing sites and length differences in the amplified sequence between primer annealing sites are some of the reasons for polymorphism. Some of the variants of RAPD include the following:

- Arbitrarily Primed Polymerase Chain Reaction (AP-PCR)—uses longer arbitrary primers than RAPDs.
- DNA Amplification Fingerprinting (DAF)—uses shorter, 5–8 bp primers to generate a larger number of fragments.
- Randomly amplified microsatellite polymorphisms (RAMPO)—genomic DNA is first amplified using arbitrary (RAPD) primers, and the amplified products are then electrophoretically separated and the dried gel is hybridized with microsatellite oligonucleotide probes.
- Multiple Arbitrary Amplicon Profiling (MAAP)—collective term for techniques using single arbitrary primers.

Advantages

- Provides a quick and efficient screen for DNA sequence-based polymorphism at a very large number of loci.
- Low quantity of template DNA requirement.
- Very high genomic abundance.
- No requirement of sequence data for primer construction.
- Does not require pre-sequencing of DNA (Bird 1978).
- Used for identification and classification of accessions (Børsting and Morling 2013), identification of breeds (Bozkurt et al. 2009), and genetic diversity analysis (Bozkurt et al. 2012).

Disadvantages

- Low reproducibility of data.
- Since they are not locus specific band profiles cannot be interpreted in terms of loci and alleles.
- Requires purified, high molecular weight DNA.
- Polymorphisms are detected only as the presence or absence of a band of a certain molecular weight, with no information on heterozygosity besides being dominantly inherited.

15.3.2.2 Amplified Fragment Length Polymorphisms (AFLP)

AFLP was developed by Peter Vos and his team in 1995 by combining the functions of RFLP and PCR. In AFLP, genomic DNA is digested using two restriction endonucleases where one is a rear cutter (Eco R1) and the other is a frequent cutter

(Mse-1). This produces sticky ends. Adapters which are double-stranded short oligonucleotide sequences of usually 14–20 base pairs are used in the sticky ends. Separate adapters are used for Eco R1 and Mse1.

The adapters are ligated and amplified using two-step PCR. The first phase of amplification is known as pre-selective amplification, and the second phase is the selective amplification. During pre-selective amplification, few fragments are selectively amplified while during selective amplification, more stringent primers (primer contain additional three nucleotides) are used to reduce the number of fragments amplified. The amplified fragments are separated using high resolution polyacrylamide gels, and the banding patterns are visualized. If separation and analysis are performed on sequencer, fluorescently labeled primers are used during selective amplification (Fig. 15.2). The variation in the length of fragments is analyzed, which gives the estimate about genetic relationships among the individuals or the variations.

Applications

- It generates fingerprints of any DNA.
- It can detect various polymorphisms in different genomic regions.
- It identifies genetic variation in strains or closely related species.
- It is used in criminal and paternity tests to determine slight differences within populations and in linkage studies to generate maps for QTL (Quantitative trait locus) analysis.
- It is used as landmarks in genetic and physical mapping since AFLP fragments correspond to unique positions on the genome.

Advantages of AFLP Marker

- It can analyze multiple locus at once.
- Less amount of genomic templates is required.
- Highly reproducible.
- No requirement of the sequence information about the organism.

Disadvantages of AFLP

- Requires high quality DNA.
- They cannot detect homozygous or heterozygous individuals since they are dominant markers.
- Since they are multi-locus, one cannot ascertain which fragment belongs to which DNA locus.

15.3.2.3 Microsatellites

The term "microsatellites" was first coined by Litt and Luty in 1989. Microsatellites are short tandem repeats of 1–6 base pairs (van Oppen et al. 2000) which has wide distribution in eukaryotes. They are also called as simple sequence repeat (SSR). They are found as di (CA), tri (AAT), tetra (GATA), penta- and hexanucleotide repeats of which the first three type of repeats are preferred for molecular genetic studies (Selkoe and Toonen 2006). They are tandemly repeated (usually 5–20 times)

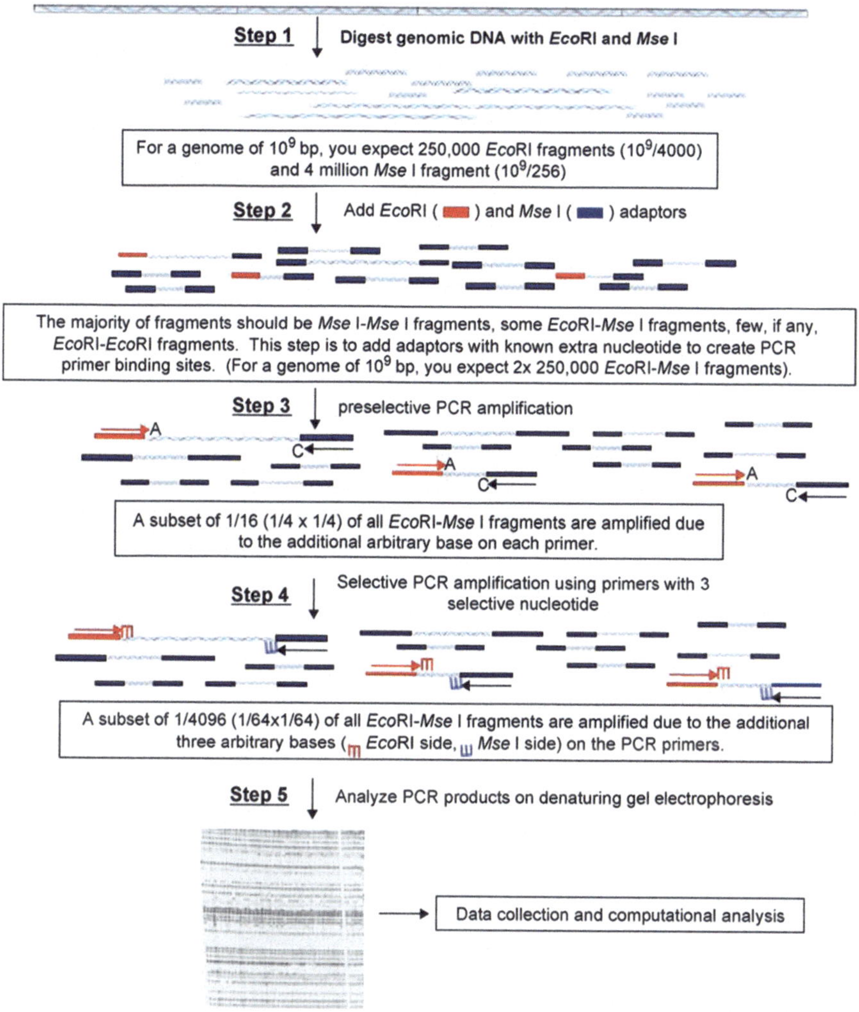

Fig. 15.2 Overview of AFLP (reproduced from Mishra et al. 2018 with permission from AkiNik Publications)

in the genome with a minimum repeat length of 12 base pairs (Goodfellow 1992; Vaughan and Lloyd 2003; Ellegren 2004). Microsatellite repeats are amplified with fluorescently labeled primers and then the alleles from each individual in a family are separated by size and the marker tested for linkage with another. Different types of tandem repeats are presented in Fig. 15.3.

The flanking regions of the microsatellite are deduced for developing them as markers. The nucleotide sequences in the flanking regions of the microsatellite must be known, in order to design specific primers to amplify the microsatellite. Micro-satellite and their flanking sequences can be identified by constructing a small-insert

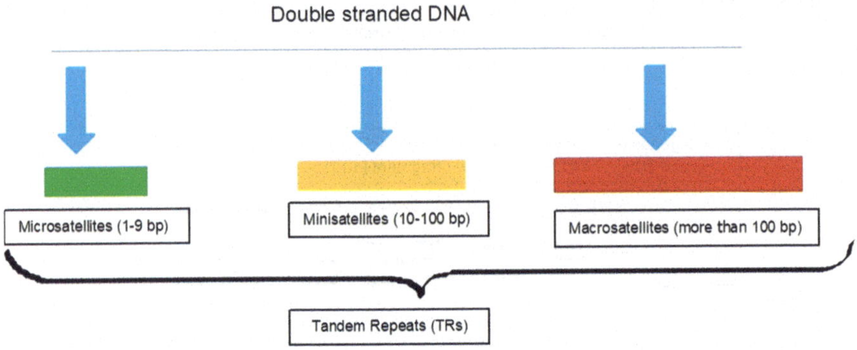

Fig. 15.3 Different types of tandem repeats

genomic library, screening the library with a synthetically labeled oligonucleotide repeat, and sequencing the positive clones. Alternatively, microsatellites may be identified by screening sequence databases for microsatellite sequence motifs from which adjacent primers may then be designed. Polymerase slippage during DNA replication, or slipped strand mispairing, is considered to be the main cause of variation in the number of repeat units of a microsatellite resulting in length polymorphisms.

Advantages of Microsatellites

- Codominance of alleles.
- High genomic abundance in eukaryotes.
- Random distribution throughout the genome.
- Low quantities of template DNA.
- Reproducibility.
- Used in detection of intervarietal polymorphisms (Stepien et al. 2007).
- Highly robust, more variable, transferable and informative than other markers (He et al. 2003; Lee et al. 2004; Senan et al. 2014).
- Screening of microsatellite variation can be automated.
- Used for parentage analysis (Senan et al. 2014).

Disadvantages of Microsatellites

- Expensive, laborious, and time-consuming (Zane et al. 2002; Squirrell et al. 2003; Thiel et al. 2003).
- Homoplasy is another problem when applying microsatellites as a reliable tool for phylogenetic analysis because alleles considered to be identical in state are not necessarily identical by descent (Estoup et al. 2002).

15.3.3 Sequence-Based Markers

The markers developed based on the sequence data of DNA are called as sequence-based markers which includes EST and SNP.

15.3.3.1 Expressed Sequence Tags (ESTs)

Expressed sequence tags are small DNA sequences of 200–800 base pairs (bp) (Fig. 15.4). They are produced by sequencing of selected cDNA clones and are stored in databases (Parkinson and Blaxter 2009). They represent the expressed DNA portion of a genome. In order to produce ESTs, mRNAs with 3′ poly-A tails are isolated and reverse transcribed to create cDNA libraries which are then cloned into vector. Individual clones from these libraries are randomly selected and subjected to a single sequencing reaction using universal primers that can be associated with either end of the insert. The raw sequence reads are processed to remove low quality sequence information, and the resultant high-quality trimmed sequence is deposited in the dbEST database through NCBI.

Advantages of ESTs

- Markers associated with coding regions of the genome can be developed.
- Polymorphic markers can be generated using the sequences.
- ESTs are superior than the random DNA based markers for marker-assisted selections.
- Using the raw sequence, the occurrence of microsatellite or single nucleotide polymorphisms can be screened.

Disadvantages of ESTs

- Recovering mRNA from some tissues are difficult.
- Information on the gene regulatory sequences present in introns are lost.
- Since they are "single pass," they have high rate of sequencing error.
- Possibility of redundancy within and between EST libraries.

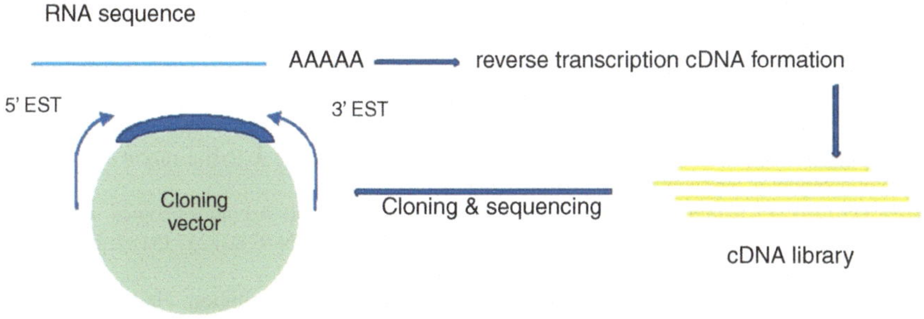

Fig. 15.4 Expressed sequence tags sequencing

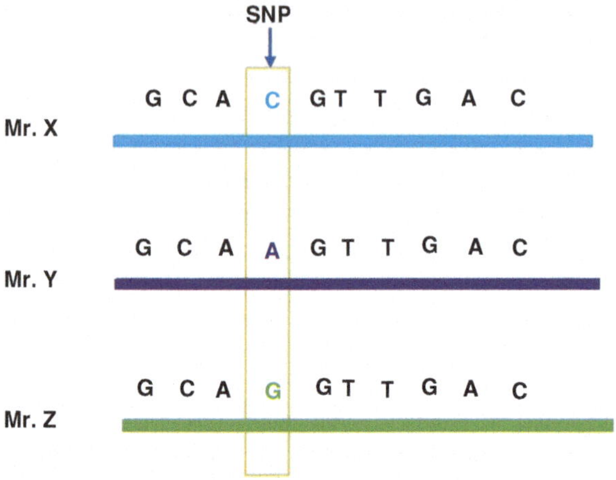

Fig. 15.5 Single nucleotide polymorphisms

15.3.3.2 Single Nucleotide Polymorphisms (SNPs)

The genomic variation at a single base point of DNA is called single nucleotide polymorphisms (Fig. 15.5) and is the reason for genetic variations. A minimum of two alleles having frequencies of more than 1% in an unrelated individuals of large population to be considered as SNP (Børsting and Morling 2013). In most cases, SNPs have only two alleles and they are more prevalent in the non-coding regions of the genome. Random shotgun sequencing, PCR amplicon targeted sequencing, RNA sequencing, and bioinformatic mining of ESTs databases are used for development and identification of SNP (Liu and Cordes 2004). They are used to construct a dense set of polymorphic markers that can be used for studying the association between the markers and a particular trait.

References

Bird AP (1978) Use of restriction enzymes to study eukaryotic DNA methylation: II. The symmetry of methylated sites supports semi-conservative copying of the methylation pattern. J Mol Biol 118:49–60

Børsting C, Morling N (2013) Single-nucleotide polymorphisms. In: Encyclopedia of forensic sciences, 2nd edn. Elsevier, pp 233–238. https://doi.org/10.1016/B978-0-12-382165-2.00042-8

Bozkurt Y, Ogretmen F, Secer FS (2009) Effect of different extenders and storage periods on motility and fertilization success of grass carp (*Ctenopharyngodon idella*) sperm during spawning season. Tarim Bilimleri Dergisi 15(3):277–284

Bozkurt Y, Yavas I, Karaca F (2012) Cryopreservation of brown trout (*Salmo trutta macrostigma*) and ornamental koi carp (*Cyprinus carpio*) sperm. In: Katkov I (ed) Current frontiers in cryopreservation. InTech Publishers, pp 293–304

Butler JM (2012) Chapter 16-non-human DNA. In: Butler JM (ed) Advanced topics in forensic DNA typing: methodology. Academic Press, pp 473–495., ISBN 9780123745132. https://doi.org/10.1016/B978-0-12-374513-2.00016-6

Ellegren H (2004) Microsatellites: simple sequences with complex evolution. Nat Rev Genet 5: 435–445

Estoup A, Jarne P, Cornuet JM (2002) Homoplasy and mutation model at microsatellite loci and their consequences for population genetics analysis. Mol Ecol 11:1591–1604

Goodfellow PN (1992) Variation is now the theme. Nature 359:777–778

He GH, Meng RH, Newman M, Gao GQ, Pittman RN, Prakash CS (2003) Microsatellites as DNA markers in cultivated peanut (Arachis hypogaea L.). BMC Plant Biol 3:3

Krawczyk B, Kur J (2018) Molecular identification and genotyping of staphylococci: genus, species, strains, clones, lineages, and interspecies exchanges. In: Savini V (ed) Pet-to-man travelling Staphylococci. Academic Press, pp 199–223., ISBN 9780128135471. https://doi.org/10.1016/B978-0-12-813547-1.00016-9

Lee JM, Nahm SH, Kim YM, Kim BD (2004) Characterization and molecular genetic mapping of microsatellite loci in pepper. Theor Appl Genet 108:619–627

Liu Z, Cordes J (2004) DNA marker technologies and their applications in aquaculture genetics. Aquaculture 238:1–37

Mishra SP, Taraphder S, Mishra C (2018) Chapter – 7, Improvement of molecular markers in animal science. In: Sreedhar S (ed) Advances in veterinary sciences, vol 2. AkiNik Publications, New Delhi, pp 95–129

Muwonge A, Egbe F, Bronsvoort M, Areda DB, Hlokwe T, Michel A (2019) Molecular epidemiology of Mycobacterium bovis in Africa. In: Dibaba A, Kriek N, Thoen C (eds) Tuberculosis in animals: an African perspective. Springer, Cham. https://doi.org/10.1007/978-3-030-18690-6_8

Parkinson J, Blaxter M (2009) Expressed sequence tags: an overview. In: Expressed sequence tags (ESTs), pp 1–12. https://doi.org/10.1007/978-1-60327-136-3_1

Selkoe KA, Toonen RJ (2006) Microsatellites for ecologists: a practical guide to using and evaluating microsatellite markers. Ecol Lett 9:615–629

Senan S, Kizhakayil D, Sasikumar B, Sheeja TE (2014) Methods for development of microsatellite markers: an overview. Not Sci Biol 6(1):1–13

Squirrell J, Hollingsworth PM, Woodhead M, Russell J, Lowe AJ, Gibby M, Powell W (2003) How much effort is required to isolate nuclear microsatellites from plants? Mol Ecol 12:1339–1348

Stepien L, Mohler V, Bocianowski J, Koczyk G (2007) Assessing genetic diversity of Polish wheat (Triticum aestivum) varieties using microsatellite markers. Genet Resour Crop Evol 54:1499–1506

Thiel T, Michalek W, Varshney RK, Graner A (2003) Exploiting EST databases for the development and characterization of gene-derived SSR-markers in barley (Hordeum vulgare L.). Theor Appl Genet 106:411–422

van Oppen MJ, Rico C, Turner GF, Hewitt GM (2000) Extensive homoplasy, non-stepwise mutations, and shared ancestral polymorphism at a complex microsatellite locus in Lake Malawi cichlids. Mol Biol Evol 17:489–498

Vaughan V, Lloyd AM (2003) An analysis of microsatellite loci in Arabidopsis thaliana: mutational dynamics and application. Genetics 165:1475–1488

Williams JGK, Kubelik AR, Livak KJ, Rafalski JA, Tingey SV (1990) DNA polymorphisms amplified by arbitrary primers are useful as genetic markers. Nucleic Acids Res 18:6531–6535

Zane L, Bargelloni L, Patarnello T (2002) Strategies for microsatellite isolation: a review. Mol Ecol 11:1–16

Biosensor

16

A biosensor is an analytical device incorporating a biorecognition element intimately associated with or integrated within a transducer that converts the biological response into an electrical signal. It is a sensor that integrates a biological element with a physiochemical transducer to produce an electronic signal proportional to a single analyte which is then conveyed to a detector. The biological response could be anything from enzyme activity or antibody/receptor binding to cell responses. The transduction to an electrical signal could also be diverse. In simple words, it is a device made by fusion of a biocomponent and an electronic gadget used to detect the presence of the substance to be monitored (Chaplin 2009).

Biosensor technologies include transduction platforms based on electrochemical (potentiometric, amperometric, impedance), piezoelectric, thermal or optical methods (reflectrometric interference spectroscopy, interferometry, optical waveguide lightmode spectroscopy, total internal reflection fluorescence, surface plasmon resonance). These techniques have been adapted to detect analytes of interest based on the interaction with or functionality modification of a biological target, which could be nucleic acids, enzymes, antibodies, receptors, cell organelles, or whole cells. The specificity of the detection is determined by the biological component of the method.

16.1 Immobilization of Biological Material

Immobilization of the biological component is the key step while constructing biosensor which can be carried out in many of the following ways as listed out by Palit and Mukherjee (2007),

- Adsorption—the biological material is contacted with a surface-active material such as graphite, alumina, clay, and glass.

C. J. Betsy, C. Siva, *Fisheries Biotechnology and Bioinformatics*,
https://doi.org/10.1007/978-981-99-6991-3_16

- Covalent attachment—chemical groups on the biological material that are not essential for activity such as non-essential amino acid residues of enzymes are attached to chemically activated supports like synthetic polymers, cellulose, and glass.
- Cross-linking—intermolecular covalent linkages can be formed between macromolecules with the aid of bifunctional cross-linking agents such as di-isothiocyanates, alkylating agents, and dialdehydes such as glutaraldehyde.
- Physical entrapment—gels or polymers such as polyacrylamides, silica gel, starch are cross-linked in the presence of enzyme or other biological material which is thereby trapped or captured.
- Microencapsulation—materials are trapped by the membranes of various polymers, usually in capsules with mean diameters ranging from 5 to 300 μm.

16.2 Components of a Biosensor

- Bio-element.
- Sensor element.
- Processor circuit.

16.3 Construction of Biosensor

Biosensor consists of biological and physical components. Biological components include enzyme, nucleic acid, antibody, etc. while physical components include transducer, amplifier, etc. The biological and physical analyte interact to produce some physical charge detectable by the transducer. As a first step, the biological component is immobilized on the transducer. Then the analyte must be transported from the solution to the biological component through simple diffusion for reaction. Then the biological component interacts specifically to the analyte which produces a physical change near the surface of the transducer. Depending on the physical change produced, the type of biosensor varies. The transducer then detects and measures this change and converts it into an electric signal. This signal is very small and is amplified by an amplifier before it is fed into the microprocessor. The signal is then processed, interpreted, and displayed in suitable units (Fig. 16.1) (Palit and Mukherjee 2007).

16.3.1 Pre-requisites for Developing Biosensor

- Selection of suitable bioreceptor molecule which requires knowledge on bio-chemistry and biology.
- Selection of suitable immobilization method which requires knowledge on chemistry.

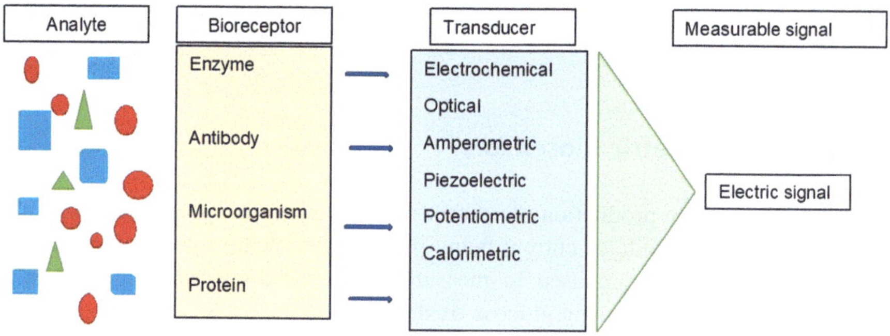

Fig. 16.1 Biosensor and its components

- Selection of suitable transducer for which knowledge on electrochemistry and physics is necessary.
- Designing a biosensor considering measurement ranger, linearity, and minimization of interference which requires knowledge on kinetics and mass transfer.
- After designing of the biosensor, it must be miniaturized and mass produced. Modern fabrication technology and micromachining technology are used in fabrication of biosensors.

16.3.2 Characteristics of Good Biosensor

- It should be highly specific for the analyte.
- The reaction used should be independent of factors such as pH, temperature, etc.
- The response should be linear over a range of analyte concentrations.
- Device should be small, durable, easy to use, and cheap.

16.4 Types of Biosensors

16.4.1 Calorimetric Biosensors

Most of the enzyme catalyzed reactions produce heat. Calorimetric biosensors measure the change in temperature of the solution containing the analyte following the enzyme action and interpret it in terms of the analyte concentration in the solution.

16.4.2 Potentiometric Biosensors

They use ion-selective electrodes that convert the biological reaction into electric signal. The electrodes employed are most commonly, glass pH electrodes for cations, glass electrodes coated with a gas selective membrane, or solid state electrodes. E.g., pH meter.

16.4.3 Amperometric Biosensors

They function by the production of current when potential is applied between two electrodes, the magnitude of current being proportional to the substrate concentration. These biosensors are used to measure redox reactions, an example being determination of glucose using glucose oxidase.

16.4.4 Optical Biosensors

These biosensors measure both catalytic and affinity reactions. They measure a change in the fluorescence or in absorbance caused by the products of catalytic reaction. E.g., biosensor using luciferase enzyme for detection of bacteria in food samples.

16.4.5 Acoustic Wave Biosensors

These are also called piezoelectric devices. Their surface is usually coated with antibodies which bind to the complimentary antigen present in the sample solution. This leads to increased mass which reduces their vibrational frequency. This change is used to determine the amount of antigen present in the sample.

16.5 Applications

- Glucose monitoring in diabetes patients.
- Environmental applications (the detection of pesticides and river water contaminants).
- Detection of pathogens.
- Determining levels of toxic substances before and after bioremediation.
- Detection and determining of organophosphate.
- Routine analytical measurement of folic acid, biotin, vitamin B12, and pantothenic acid as an alternative to microbiological assay.
- Determination of drug residues in food, such as antibiotics and growth promoters, particularly meat and honey.
- Drug discovery and evaluation of biological activity of new compounds.
- Detection of toxic metabolites such as mycotoxins.

16.6 Applications in Fisheries

16.6.1 Biosensors in Food Analysis

Biosensors in food industry uses optic coated with antibodies to detect pathogens and food toxins. The light system in these biosensors has fluorescence. A range of immune-and ligand-binding assays for the detection and measurement of small molecules such as water-soluble vitamins and chemical contaminations (drug residues) such as sulfonamides and beta-agonists has been developed.

16.6.2 Biosensors in Aquatic Environment Assays

Due to water pollution, fish has shown reproductive dysfunction with males displaying feminization. The biochemical responses of organisms to organic and metal compounds in the water can be measured and used as a biomarker to study the level of pollution. Most commonly, Cytochrome P4501A is used as it is responsive to a number of organic chemicals including aromatic hydrocarbons and dioxins. The induction of this gene by these contaminants is measured by changes in protein expression or mRNA levels. Alternatively, metallothionein is utilized, which are induced specifically by metals.

Similarly, gene encoding green fluorescent protein (GFP) is fused to a number of promoters which will respond to water pollutants. These include the promoters from some inducible genes such as those encoding heat shock proteins or metallothioneins which are induced by general stress, heavy metals, or chemical toxins, those contain estrogen response elements being induced by estrogens or xenoestrogens. The availability of the GFP as a reporter gene has enabled the use of transgenic organisms as biosensors for water contamination providing rapid and visible results while eliminating the need for enzymatic or specific protein assays.

16.6.3 Aequorin

Aequorin is a photoprotein isolated from the bioluminescent jellyfish (*Aequorea victoria*). Upon addition of calcium ions (Ca^{++}) and coelenterazine, a reaction occurs whose end result is the generation of blue light in the 460–470 nm range. Aequorin has been incorporated into human B cell lines for the detection of pathogenic bacteria and viruses in the CANARY assay (Cellular Analysis and Notification of Antigen Risks and Yields). The B cells are genetically engineered to produce aequorin. Upon exposure to antigens of different pathogens, the recombinant B cells emit light as a result of activation of an intracellular signaling cascade that releases Ca^{++} inside the cell.

16.6.4 Green Fluorescent Protein (GFP)

Green fluorescent protein (GFP) is also a photo protein isolated and cloned from the jellyfish *A. victoria*. GFP, like aequorin, produces a blue fluorescent signal, but without the required addition of an exogenous substrate. All that is required is an UV source to activate the fluorescent properties of the photoprotein. This ability to autofluorescence makes GFP highly desirable in biosensing assays since it can be used on line and in real time to monitor intact, living cells. Additionally, the ability to alter GFP to produce light emissions besides blue (i.e., cyan, red, and yellow) allows it to be used as a multianalyte detector. Consequently, GFP has been used extensively in bioreporter constructs within bacterial, yeast, nematode, plant, and mammalian hosts.

16.6.5 Biosensors in the Detection of Fish Toxins

Biosensors technologies can detect different toxins in seafood. Several biosensor techniques have been developed that allow the detection of tetrodotoxin. One of the approaches uses an anti-tetrodotoxin specific antibody as the biological reporter and amperometric detection with a screen-printed electrode. The format of the assay is an indirect competition assay where the amount of current generated by *p*-aminophenol, the product of the enzymatic activity of the alkaline phosphatase label of the specific antibody, is inhibited by the presence of tetrodotoxin in a sample that competes with the electrode immobilized tetrodotoxin.

Another tetrodotoxin biosensor based on inhibition of cell function has been designed using murine spinal cord neuronal networks cultured on microelectrode arrays. The biological response is monitored as extracellular potentials. In this system, tetrodotoxin quantification was based on spike rate inhibition.

16.6.6 Lab-on-a-Fish from PNNL

Pacific Northwest National Laboratory has developed a biosensor named "Lab-on-a-fish" that can collect data about a fish, including its location, heartbeat, tail movement, and burned calories, as well as the temperature, pressure, and magnetic field of its surrounding environment. It also uses electrocardiogram and electromyogram. This will help to understand the impact of climate change and infrastructure development on ecosystem health and develop suitable management and conservation strategies. This particular device was developed in order to create a better fish passage at hydraulic structures.

16.6.7 BIOLAN for Sulfite Testing

BIOLAN has developed a portable biosensor for sulfite monitoring in shrimp farming. It requires minimal sample and is highly suitable for field applications to ensure the right amount of metabisulfite in water. The optimized biosensor provides accurate sulfite quantification from 8 to 110 g/L of metabisulfite, a repetitivity expressed as an imprecision lower than 7%, and as an irreproducibility lower than 10%.

References

Chaplin MF (2009) Biosensors. In: Molecular biology and biotechnology, 5th edn. RSC Publishing, Cambridge, pp 513–547

Palit D, Mukherjee A (2007) Biosensors: an appraisal of its operation and applications. In: Environmental biotechnology. APH Publishing Corporation, New Delhi, pp 55–82

17 Bioremediation

Bioremediation is the process whereby organic wastes are biologically degraded under controlled conditions to levels below concentration limits established by regulatory authorities. It uses naturally occurring microorganisms like bacteria and fungi or plants to degrade or detoxify substances hazardous to human health and/or the environment.

17.1 Principle of Bioremediation

Bioremediation allows natural processes to clean up harmful chemicals in the environment. Microbes that live in soil and groundwater consume certain harmful chemicals to enhance their growth and metabolism. The microorganisms may be indigenous to a contaminated area or they may be isolated from elsewhere and brought to the contaminated site. Contaminant compounds are transformed by living organisms through reactions that take place as a part of their metabolic processes. Biodegradation of a compound is often a result of the actions of multiple organisms. Microorganisms must enzymatically attack the pollutants into digestible pieces. When microbes completely digest these chemicals, they change them into water and harmless gases such as carbon dioxide. Bioremediation can be effective only where environmental conditions permit microbial growth and activity.

17.1.1 Factors Determining the Success of Bioremediation

- Existence of a microbial population.
- Availability of contaminants to the microbial population.
- Environmental factors (type of soil, temperature, pH, the presence of oxygen, nutrients).

C. J. Betsy, C. Siva, *Fisheries Biotechnology and Bioinformatics*,
https://doi.org/10.1007/978-981-99-6991-3_17

17.1.1.1 Microbial Populations

Microorganisms can be isolated from almost any environmental conditions. Microbes will adapt and grow at subzero temperatures, as well as extreme heat, desert conditions, in water, with an excess of oxygen, and in anaerobic conditions, with the presence of hazardous compounds or on any waste stream. They require only an energy source and a carbon source.

17.1.1.2 Bio-stimulation

Bio-stimulation is the addition of nutrients and oxygen to help indigenous microorganisms. Nutrients allow microbes to create the necessary enzymes to break down the contaminants. Hence all microbes need nitrogen, phosphorus, and carbon. Carbon is the most basic element of living forms and is needed in greater quantities than other elements. The nutritional requirement of carbon to nitrogen ratio is 10:1, and carbon to phosphorous is 30:1. For degradation, it is necessary that bacteria and the contaminants be in contact. This is not easily achieved, as neither the microbes nor contaminants are uniformly spread in the soil. Some bacteria are mobile and exhibit a chemotactic response, sensing the contaminant and moving toward it. Other microbes such as fungi grow in a filamentous form toward the contaminant. It is possible to enhance the mobilization of the contaminant utilizing some surfactants such as sodium dodecyl sulfate.

17.2 Bioremediation in Fisheries

17.2.1 Bioremediation Using Algal Mat

Microbial mats constitute laminated heterotrophic and autotrophic bacterial communities that include cyanobacteria, diatoms, an oxygenic phototrophic bacteria, and sulfate reducing bacteria (D'amellio et al. 1989). These mats can be cultured on large scale to produce protein via nitrogen fixation in aquaculture. These mats are known to sequester or precipitate metals/radionuclides by surface absorption, thereby bioconcentrating the metal to small volume. Organic contaminants can be completely degraded and mineralized.

17.2.2 Bioremediation Using Bagasse

In wastewater from shrimp farms, substrate based on bagasse is known to remove total ammoniacal nitrogen (TAN) by 95% in 144 h. Removal of TAN using enriched immobilized cultures of autochthonous nitrifying bacteria is reported to have advantages such as longer biomass retention in pond water, maintenance of high microbial cell density, optimization of microbial growth and metabolic rates, and protection from inhibitory compounds (Travieso et al. 1995).

17.2.3 Bioremediation of Wastewater Using Aquatic Weeds

Use of water hyacinth, *Eichhornia crassipes* and other aquatic weeds such as *Spirodela*, *Lemna*, and *Wolffia* has been recommended for treating wastewater. These species are reported to absorb and accumulate the heavy metals and other toxic elements such as sulfur, iron, potassium, etc. *E. crassipes* especially can absorb phenols, benzene, and toxaphene and certain radioactive toxicants which are difficult to remove. These weeds can be harvested periodically and useful minerals can be extracted from them and used for biogas production and as fertilizer (Mishra and Juwarkar 2007).

17.2.4 Bioremediation of Wastewater Using Earthworms

It was reported that earthworms can remove pollutants from water by filtering the water through vermin-filter. The earthworms convert the impurities into worm-casts which can be utilized as fertilizer as they are rich in beneficial microbes and organic nutrients. The treated water can be used for gardening (Sharma 2001).

17.2.5 Bioremediation of Spilled Oil

Oil spills on seas are considered as a major environmental hazard as they destroy the habitats of aquatic animals. Due to the rapid development in technology, a genetic engineered oil utilizing bacteria that grow rapidly on oil was produced. *Pseudomonas aeruginosa* produces a glycolipid emulsifier that reduces the surface tension of oil water interface and removes oil from water (Jogdan 1995; Gupta 1996).

17.2.6 Bioremediation of Organophosphorus Pesticides

Apart from agriculture, pesticides are widely used in medical sector for controlling mosquitoes and rodents, in urban area for termite control, in industrial, commercial, municipal, and military sectors for rodent and weed control. These pesticides enter into water through runoff and pollute the aquatic environment and degrade the habitats and leads to mortality in severe cases. These pesticides are also reported from groundwater sources with traces of cadmium, fluoride, arsenic, nitrates, and lead. Presence of arsenic in drinking water can lead to cancer and other severe illnesses. Some of the important organophosphorus pesticides include parathion, methyl parathion, aminoparathion, fenitrothion, diazinon, and coumaphos. Apart from these, *p*-Nitrophenol (PNP), the metabolite of parathion and methyl parathion and 3-methyl-4-nitrophenol, the metabolite of fenitrothion are also important. Microorganism such as *Flavobacterium* sp. (Sethunathan and Yoshida 1973), *Pseudomonas stutzeri* (Daughton and Hsieh 1977), and *Bacillus* sp. (Siddharamappa et al. 1973) are effective against PNP. *Bacillus subtilis*, *B. cereus*, *B. megaterium*,

B. macerans, *Alcaligenes viscolactis*, *Serratia plymuthica* (Miyamoto et al. 1966; Hirakoso 1968), *Rhizobium japonicum*, and *R. meliloti* (Mick and Dahm 1970) can also be used to treat organophosphorus pesticides.

17.2.7 Other Bioremediation

In aquaculture, organic detritus are removed using *Bacillus* sp. and *Lactobacillus* sp., while nitrogenous compounds can be removed by *Nitrosomonas*, *Nitrobacter*, *and Nitrococcus*. It is reported that hydrogen sulfide can be removed by using species belonging to Chromatiaceae and Chlorobiaceae.

17.3 Advantages of Bioremediation

- Bioremediation is perceived by the public as an acceptable waste treatment process.
- It is safe as the residues for the treatment are usually harmless products and include carbon dioxide, water, and cell biomass.
- It is useful for the complete destruction of a wide variety of contaminants. This eliminates the chance of future liability associated with treatment and disposal of contaminated material.
- Instead of transferring contaminants from one environmental medium to another, for example, from land to water or air, the complete destruction of target pollutants is possible.
- It can often be carried out on site, without disruption of normal activities, no need to transport waste off site.
- It does not require sophisticated equipment.
- Less expensive.

17.4 Disadvantages of Bioremediation

- Not all compounds are susceptible to rapid and complete degradation.
- Biological processes are often highly specific.
- Important site factors required for success include the presence of metabolically capable microbial populations, suitable environmental growth conditions, and appropriate levels of nutrients and contaminants.
- Research is needed to develop and engineer bioremediation technologies for complex mixtures of contaminants that are not evenly dispersed in the environment.

References

D'amellio ED, Cohen Y, Des Marias D (1989) Comparative functional ultrastructure of two hypersaline submerged cyanobacterial mats. In: Microbial mats. American Society for Microbiology, Washington, pp 97–113

Daughton CG, Hsieh DPH (1977) Accelerated parathion degradation in soil by inoculation with parathion-utilizing bacteria. Bull Environ Contam Toxicol 18:48–56

Gupta PK (1996) Elements of biotechnology. Part IV biotechnology and environment. Vivek Rastogi Company, Meerut, p 602

Hirakoso S (1968) World Health Organisation-WHO/VBC, vol 68, pp 1–4

Jogdan SN (1995) Environmental biotechnology, (industrial pollution management). Modern College, Vashi, Himalaya Publishing House, New Bombay

Mick KL, Dahm PA (1970) J Econ Entomol 63:1151–1159

Mishra CSK, Juwarkar AA (2007) Environmental biotechnology. APH Publishing Corporation, New Delhi, p 466

Miyamoto J, Kitagawa K, Sato Y (1966) Metabolism of organophosphorus insecticides by Bacillus subtilis with special emphasis on Sumithion. Jpn J Exp Med 36:211–225

Sethunathan N, Yoshida T (1973) A Flavobacterium sp. that degrades diazinon and parathion. Can J Microbiol 19:873–875

Sharma BK (2001) Environmental chemistry. Goel Publishing, Meerut

Siddharamappa R, Rajaram KP, Sethunathan N (1973) Degradation of parathion by bacteria isolated from flooded soil. Appl Microbiol 26:846–849

Travieso L, Benitez F, Weiland P, Sanchez E, Dupeyron R, Dominguez AR (1995) Experiments on immobilization of microalgae for nutrient removal in wastewater treatment. Bioresour Technol 55:181–186

Bioprospecting in Fisheries 18

Bioprospecting, also known as biodiversity prospecting, is the exploration of biological material for commercially valuable genetic and biochemical properties. It is the investigation of living things to see how they can be commercially useful to humans. Small samples of natural resources are collected for their potential value to industry, particularly in the pharmaceutical, biotechnology, and agribusiness fields.

Local communities close to where the material originates may have specialized knowledge on how the resources are used, which can also be collected, and this is known as traditional or indigenous knowledge (IK). When biodiversity or related knowledge is collected without permission from the owners of these resources and then patented, it is known as biopiracy.

Bioprospecting cuts across a range of different sectors, including pharmaceutical, agribusiness (biotechnology, seed, crop protection, and horticulture), cosmetic and personal care, fragrance, botanicals, and the food and beverage industries. Degree and way in which these sectors access and use these resources vary significantly. Pharmaceutical and agribusiness sectors are more involved in this. Cosmetic, fragrance, botanicals, and food and beverage industries still do bioprospecting, but are not always aware of, or compliant to, the regulation and legislation and may commit biopiracy by commercializing natural resources.

In the past 15–20 years, large pharmaceutical companies closed their natural products programmes. This is due to the "slim pickings" and extensive financial outlay required, combined with complex negotiations over intellectual property (IP). Natural products research is very resource intensive requiring lots of money, people, and expertise, making large companies reluctant to get involved again significantly. Smaller, focused companies are becoming productive by obtaining leads and "hits" and then working with the larger companies to develop the products. Today, the majority of natural products research is undertaken in academic and government research institutes or in smaller discovery companies.

C. J. Betsy, C. Siva, *Fisheries Biotechnology and Bioinformatics*,
https://doi.org/10.1007/978-981-99-6991-3_18

18.1 Phases Involved

- Collecting.
- Analysis.
- Commercialization.
- Samples and/or indigenous knowledge related to a sample are collected, and then undergo analysis—using a variety of different technologies.

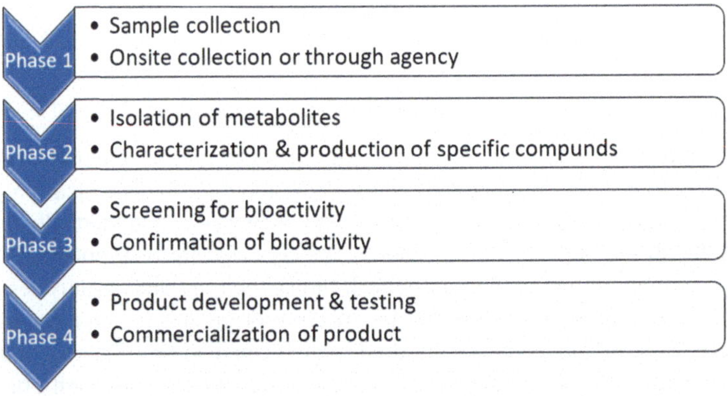

Once a "hit" has been identified, it can take 20 years (depending on how the material is used) before the final product completes clinical trials and can be commercialized. It can take 12–15 years and upward of US$800 million indirect and non-direct costs to bring a drug to the US market. Only 20% of drugs that begin clinical testing proceed to trial and eventually for marketing approval. Only three out of ten drugs that are finally marketed recoup their development costs. It is a long and costly process which may result in little or no benefits.

18.2 Bioprospecting Marine Biological Resources

Marine bioprospecting is a targeted and systematic search in marine organisms including bacteria, fungi, and viruses and larger organisms such as sea plants, shellfish, and fish from the sea, the fjords, seabed, or from oil reservoirs beneath the seabed. The result of the bioprospecting could be a purified molecule that is produced biologically or synthetically, or the entire organism. The purpose of marine bioprospecting is to find components, compounds, or genes that may be included as components in products or processes that have application in medicine, processing industries, including oil and gas, food, feed, and biofuels.

Antimicrobial, cytotoxic, anti-tumor, tumor-promoting, anti-inflammatory, and antiviral agents are isolated from sessile, soft-bodied marine organisms. Molluscan metabolites are tested for neuromuscular blocking action, anti-predator, antimicrobial, anti-neoplastic, and cytotoxic activity. Dolastatin 10 is an anti-neoplastic peptide isolated from the sea hare *Dolabella auricularia*. The medical properties of this sea hare were known to the ancient Greeks and Romans, with the extracts being recommended for the treatment of some diseases as early as 200 BC. Dolastatin 10 has recently reached clinical trials in the United States and is reported to be one of the most potent anticancer agents known. Sea hare, *Aplysia* spp. produces a substance that shows good potential for curing bladder cancer in human beings.

Toxic peptides produced by predatory snails in the genus *Conus* are used as tools for neurological research and traded as strong pain killers and potential drugs against epilepsy, depression, and schizophrenia. Ziconotide occurs along with other peptides in the venom of the predatory Indo-Pacific marine molluscs, *Conus magus. C. magus* and other *Conus* species are fish-hunting molluscs that use their venom to paralyze their prey.

Nerve toxin, conotoxins, block channels regulating the flow of potassium or sodium across the membranes of nerve or muscle cells are valuable probes in physiological and pharmacological studies. Turbinaria is a genus of brown algae (Phaeophyceae) found primarily in tropical marine waters—used as a way of removing lead from aqueous solutions. *Padina pavonica* is a brown alga of the family Phaeophyceae, and active molecules of the extract of *P. pavonica* amplify the fixation of calcium by osteoblasts.

Horseshoe crabs produce copper-based blue blood containing a substance called Limulus amebocyte lysate which coagulates in the presence of small amounts of bacterial toxins and is used to test for sterility of medical equipment and virtually all intravenous drugs. Marine fish derived bioactive peptides can be developed as antihypertensive components in functional foods, or nutraceuticals are used as potential drug candidates for preventing and treating high blood pressure.

Marine organisms are known to have a better hit-rate in anti-tumor activities (Fernando de la Calle 2009). It has been estimated by the US National Cancer Institute that 1% of samples from marine animals tested in the laboratory reveal anti-tumor potential, compared with just 0.01% of samples from terrestrial origin (Fernando de la Calle 2009). One of the chemical compounds derived from marine species is conotoxin in cone snail venoms, which normally could kill humans but may be used as drugs for the treatment of epilepsy, cardiovascular disease, and psychiatric disorders. The first conotoxin-based medicine, Prialt, was approved by the US Food and Drug Administration (FDA) in December 2004 and is in the market (David Leary et al. 2009). The details of marine organisms with pharmaceutical properties and the compounds derived from it are presented in Tables 18.1 and 18.2, respectively.

Table 18.1 Marine organisms with pharmaceutical properties

Organism	Properties of the compound
Marine sponges, *Hemiastrella minor*	Anti–neoplastic
Corals, *Sinularia polydactyla*	Anti-tumor
Marine sponges, *Discodermia dissoluta*	Immune suppressive & antimitotic
Marine sponges, *Petrosia contignata*	Antiasthma
Deep sea microbes, *Thermus thermophilus*	Antiaging
Marine cyanobacteria, *Lyngbya majuscula*	Chemotherapy
Marine cyanobacteria, *Tolypothrix distorta*	Antifungal
Jelly fish, *Aequorea victoria*	Bone repair
Cone snails, *Conus magus*	Neuropathic pain
Marine sponges, *Halichondria okadai*	Anti-microtubule
Marine fungi, *Aspergillus* sp. Mangrove tunicate, *Ecteinascidia turbinata* marine actinomycetes, *Salinispora tropica* Green algae, *Dunaliella tertiolecta* Diatom, *Skeletonema marinoi* Microalga, *Isochrysis galbana* Marine sponges, *Krikpatrika variosa*	Anticancer
Soft coral, *Pseudopterogorgia elisabethae* Diatom, *Cylindrotheca closterium*	Anti-inflammatory
Brown seaweeds, *Fucus vesiculosus*	Neuro protective
Marine fungi, *Acremonium chrysogenum*	Antibacterial
Marine kelp, *Ascophyllum nodosum*	Antidiabetic
Bacterium, *Nocardiopsis dassonvillei*	Antimicrobial, antioxidant, anti-inflammatory
Nannochloris atomus	C fatty acids
Leptolyngbya	Phycocyanin
Microalga, *Tetraselmis* sp.	β-carotene, neoxanthin, violaxanthin, and lutein production
Microalga, *Isochrysis galbana*	Fucoxanthin, β-carotene production
Cyanobacterium, *Nostoc* sp., azolla-anabaena	Nitrogen-fixing biofertilizers
Diatom, *Cyclotella cryptica; Mayamaea* sp.	Lipid production
Yeast, *Saccharomyces cerevisiae*	Bioethanol production
Bacterium, *Alteromonas macleodii*	Exopolysaccharides production
Brown alga, *Saccharina japonica*	Fucoidan production
Archea, *Desulfurococcus* sp., *Pyrococcus* sp., *Thermococcus* sp.	Enzymatic bioremediation
Bacterium, *Burkholderia sacchari*	Enzymatic production of bioplastics
Fungus, *Aspergillus* sp.	Xylanolytic activity

Table 18.2 Pharmaceutical compounds derived from marine organisms

Organism	Compound derived
Marine sponges, *Hemiastrella minor*	Jaspamide
Corals, *Sinularia polydactyla*	Cembranoid diterpenes
Marine sponges, *Krikpatrika variosa*	Variolin
Marine sponges, *Discodermia dissoluta*	Discodermolide
Marine sponges, *Petrosia contignata*	Contignasterol
Deep sea microbes, *Thermus thermophilus*	Venuceane
Marine cyanobacteria, *Lyngbya majuscula*	Cryptophycin
Marine cyanobacteria, *Tolypothrix distorta*	Tolytoxin
Jelly fish, *Aequorea victoria*	Aequorin
Mangrove tunicate, *Ecteinascidia turbinata*	Trabectedin
Cone snails, *Conus magus*	Prialt
Marine sponges, *Halichondria okadai*	Eribulin
Marine fungi, *Aspergillus* sp.	Plinabulin
Soft coral, *Pseudopterogorgia elisabethae*	Pseudopterosins
Marine actinomycetes, *Salinispora tropica*	Salinosporamide A
Green algae, *Dunaliella tertiolecta*	Violaxanthin
Brown seaweeds, *Fucus vesiculosus*	Fucoidan
Marine fungi, *Acremonium chrysogenum*	Cephalosporin P
Marine kelp, *Ascophyllum nodosum*	Phlorotannins

18.3 Regulation of Bioprospecting

The politics of bioprospecting are highly complex and continue to hinder natural product development. The most significant change and the starting point of international regulation and legislation took place at the Rio Earth Summit in 1992, when participating countries discussed bioprospecting and signed the Convention of Biological Diversity (CBD). CBD recognizes the need for conservation, sustainable use, and equitable benefit sharing as cornerstones, and for the first time, acknowledged the sovereign rights of countries over their natural resources. It meant that biodiversity could no longer be regarded as common property and that the origin of the material used was recognized.

18.3.1 Bonn Guidelines (Sirakaya 2019)

- Access to genetic resources.
- Specialized access and benefit sharing regimes consistent with the objectives of the CBD.
- Meet the ethical concerns, competing requirements between freedom of scientific research.

- Exploitation of a resource and benefit to the source owner and society at large.
- Use of traditional knowledge associated with genetic resources within the scope.

18.3.2 United Nations Convention on the Law of the Seas (UNCLOS) (Guilfoyle 2019)

- An international agreement that establishes guidelines for the exploitation of marine resources and limiting the national rights of a country in the world's ocean.
- Defines the rights and responsibilities of nations with respect to their use of the world's oceans.
- Establishing guidelines for businesses, the environment, and the management of marine natural resources.

18.3.3 Global Ocean Commission (Global Ocean Commission 2014)

- International entitative aiming to reduce the degradation of the marine environment with a special focus on seas beyond the exclusive economic zones.
- Recommend policies for governance in the high seas.
- Recommend amendment of UNCLOS.
- Enable governance at the regional level.

18.3.4 International Sea Bed Authority (Jaeckel 2020)

- Established by UNCLOS.
- Organization established by the UNCLOS, aiming to regulate the prospecting and exploitation of marine minerals in international seabed areas.

18.3.5 Valencia Declaration (Laladhas et al. 2016)

- Governance regime for the regulation of activities in the marine areas.
- Urging international regulation of marine activities beyond national jurisdictions to ensure the balance between human's benefit and protection of marine biodiversity.

18.3.6 European Micro B3 (Chege kamau et al. 2015)

- Development of standards for sampling of marine microorganism.
- Sets a legal framework for creating a data base containing the genomes and metagenomes of marine microorganisms.

18.3.7 European Science Foundation

- Envisages to develop principles on the simplification and harmonization of regulations on access and fair and equitable benefit sharing arising from the exploitation of marine genetic resources.

18.3.8 Nagoya Protocol (Martins et al. 2020)

- Supplementary agreement to the CBD, promoting access and benefit sharing of genetic resources.
- Adopted in October 2010 and entered into force in October 2014.

18.4 Biopiracy

There exists inequality between developed countries and developing countries that are rich in biological resources. Developed countries generally own the technology to conduct bioprospecting by utilizing the resources from biodiversity-rich developing countries. In return, developing countries have demanded developed countries to comply with the regime of access and benefit sharing under relevant international instruments, especially the CBD. The biodiversity-rich developing countries in general have argued that there has been an imbalance and even injustice in the sharing of wealth generated from the utilization of genetic resources. This has led to the coining of the term "biopiracy" in 1993 by the Rural Agricultural Foundation International (RAFI), a Canada-based non-governmental organization, which is currently known as the Action Group on Erosion, Technology and Concentration (ETC). Biopiracy itself can be generally defined as "the illegitimate appropriation or commercialisation of genetic resources and associated knowledge" (Alexiades and Laird 2002). There are three categories of biopiracy, namely patent-based biopiracy, non-patent biopiracy, and misappropriations (Robinson 2010).

- Patent-based biopiracy—patenting of inventions based on biological resources and/or traditional knowledge without adequate authorization and benefit sharing from other countries, indigenous, or local communities.
- Nonpatent biopiracy—other intellectual property control, such as plant variety protection or trademarks.
- Misappropriations—unauthorized use of biological resources and/or traditional knowledge from other countries, indigenous, or local communities, without adequate benefit sharing.

References

Alexiades MN, Laird SA (2002) Laying the foundation: equitable biodiversity research relationships. In: Laird SA (ed) Biodiversity and traditional knowledge: equitable partnerships in practice, vol 3. Earthscan, p 7

Chege Kamau E, Winter G, Stoll PT (2015) 20 Micro B3 model agreement on access to marine microorganisms and benefit sharing. In: Research and development on genetic resources: public domain approaches in implementing the Nagoya protocol, 1st edn. Routledge

de la Calle F (2009) Marine genetic resources: a source of new drugs-the experience of the biotechnology sector. Int J Mar Coast Law 24:209. 210

Global Ocean Commission (2014) From decline to recovery: a rescue package for the global ocean

Guilfoyle D (2019) Oceans governance, the UN convention on the law of the sea and its implementing agreements. SSRN J:1–43

Jaeckel A (2020) Strategic environmental planning for deep seabed mining in the area. Mar Policy 114:103423

Laladhas KP, Nilayangode P, Oommen OV (2016) Biodiversity for sustainable development. Springer

Leary D et al (2009) Marine genetic resources: a review of scientific and commercial interest. Mar Policy 33:183–185

Martins J, Cruz D, Vasconcelos V (2020) The Nagoya protocol and its implications on the EU Atlantic area countries. J Mar Sci Eng 8:92

Robinson DF (2010) Confronting biopiracy: challenges, cases and international debates. Earthscan, p 21

Sirakaya A (2019) Balanced options for access and benefit-sharing: stakeholder insights on provider country legislation. Front Plant Sci 10:1175

Bioprocess Engineering in Fisheries 19

It deals with the design and development of equipment and processes for the manufacturing of products such as agriculture, food, feed, pharmaceuticals, nutraceuticals, chemicals and polymers, and paper from biological material and treatment of waste water. It's a combination of knowledge in biotechnology and engineering. Bioprocess engineering emphasizes the engineering and sciences of industrial processes that are biobased:

- Biomass feedstock conversion for a sustainable society or biorefinery
- Biocatalysis-based processing
- Manipulation of microorganisms for a sustainable and socially desirable goal.

19.1 Stages Involved

- Upstream processing
- Reactor engineering
- Downstream processing

19.1.1 Upstream Processing

The upstream processing is the initial step in which the microorganisms or cell lines are grown in bioreactors which include inoculum preparation, media formulation, improvement of inoculum by genetic engineering, growth kinetics optimization with an aim to increase the production levels along with the regulation of temperature, pH, and pressure (Fig. 19.1). It includes selection of a microbial strain characterized by the ability to synthesize a specific product having the desired commercial value. The strain then is subjected to improvement protocols to maximize the ability of the strain to synthesize economical amounts of the product.

C. J. Betsy, C. Siva, *Fisheries Biotechnology and Bioinformatics*,
https://doi.org/10.1007/978-981-99-6991-3_19

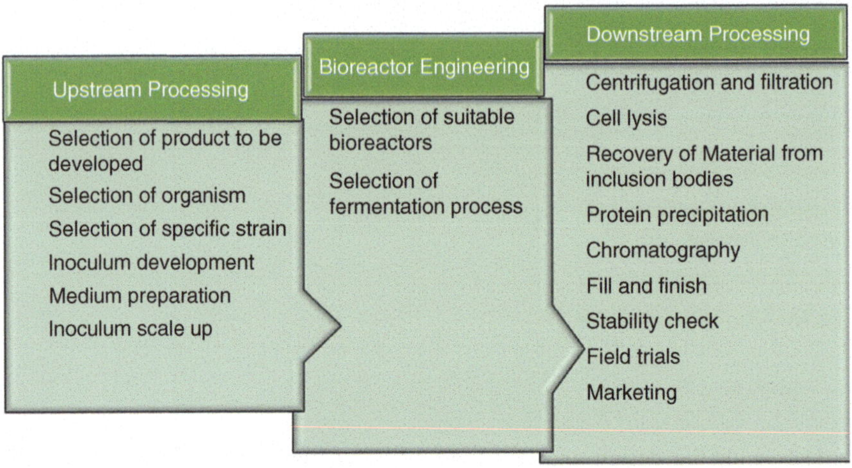

Fig. 19.1 Process of bioprocess engineering

19.1.1.1 Inoculum Development

Inoculum development refers to the increase in the density of viable number of microorganisms from a dormant stock stage to a final productive stage. Inoculum media are usually balanced for rapid cell growth and not for product formation. The criteria to be considered when preparing a protocol for inoculum development are given below.

1. Active and healthy inoculum is required to reduce the lag phase.
2. Sufficient large volumes required with an optimum size of 0.1–10% of the medium volume.
3. Suitable morphological form.
4. Contamination free.
5. Possess product forming capabilities.

19.1.1.2 Medium Preparation

All microorganisms require carbon, nitrogen, vitamins, minerals, water, and oxygen for their growth. Hence, the formulation of the nutrients is the most important requisite for the optimized production of the desired product, cell biomass, or a specific metabolite. Media formulations and environmental conditions drastically affect physiology, biochemistry, and the morphological changes which further influence product yield and other fermentation properties. First strategy for a rational design of fermentation media is according to the elementary composition of a specific microorganism and of the main product that is produced. Second is to design according to the stoichiometry of growth and product formation which is represented by the chemical equation as:

$$C_wH_xO_yN_z + aO_2 + bH_gO_hN_i \xrightarrow[C_jH_kO_lN_m]{} cCH_\alpha O_\beta N_\delta + dCO_2 + eH_2O + f$$

(Substrate) (Nitrogen source) (Biomass)

(Product)

Criteria for Selection of a Medium

- Maximum product or biomass yield per gram substrate consumed.
- Maximum purity of product or biomass.
- Maximum product formation rate.
- Low unwanted product yield.
- Cheap, consistent quality, and availability throughout the year.
- Minimal troubleshoots in upstream and downstream processing.

19.1.1.3 Criteria for Selection of Raw Materials

- Inexpensive, consistent quality, and easy availability.
- Ease of transport in any form.
- Withstand sterilization procedures.
- Favorable properties that may influence agitation, aeration, and foaming.
- Increased levels of desired product with respect to its production rate and yield per gram of substrate.
- Low quantities of impurities.
- Overall health and safety implications.

19.1.1.4 Inoculum Scale Up

It is scaling up of the medium to be used in the larger fermenter vessel. It involves growing the microorganisms obtained from the pure stock culture in several consecutive fermenter. This reduces the time required for the growth of microbes in the fermenter and increases the rate of productivity. The seed culture obtained is then used for inoculation in fermentation medium. The size of the inoculum is generally 1–10% of the total volume of the medium.

The aim of scale-up is to develop optimal environmental and operating conditions at different levels for a successful fermentation industry where conditions like substrate concentration, agitation and mixing, aeration, power consumption, and rate of oxygen transfer are studied. In a conventional scale-up, a fermentation technique is developed in three to four stages. The initial stage involves a screening process using Petri dishes or Erlenmeyer flasks followed by a pilot project to determine the optimal operating conditions for a fermentation process with a capacity of 5–200 L. The final stage involves the transfer of technology developed in the laboratory to industry.

19.1.1.5 Preparation and Sterilization of Media

Medium components should be free from contamination which requires sterilization of all the medium components. Sterilization is mostly carried out by applying heat and to lesser extent other physical methods, chemical methods (disinfectants), and radiation (using UV rays, γ rays). Batch sterilization is carried out at 121 °C (20–60 min), whereas continuous sterilization is done at 140 °C for 30–120 s. Air and heat sensitive components are sterilized by membrane filters.

19.1.2 Reactor Engineering

Fermentation process is carried out in large tanks known as fermenters or bioreactors. In addition to mechanical parts which provide proper conditions inside the tank such as aeration, cooling, agitation, etc. the tank is usually also equipped with complex sets of monitors and control devices in order to run the microbial growth and product synthesis under optimized conditions. The processing of the fermentation reactions inside the fermenter can be done using many modifications of engineering technologies.

19.1.2.1 Fermentors

Fermentors also called as bioreactors are a system which provide suitable environment for organism to efficiently produce target product. In other words, fermentor is a place where fermentation takes place. The first fermentor was developed by Chain Weizmann for the production of acetone between 1914 and 1915. Later in 1930, large scale aerobic fermentor was developed. Following this in 1934, Strauch and Schmidt introduced aeration system, water and steam inlet flow in a fermentor which are essential components of the system. Fermentor for yeast production was developed by De Beeze and Liebmann during 1944. In general, fermentors are used to produce metabolites, enzymes, vaccines, therapeutic proteins, microbial biomass, and recombinant product (Rehm et al. 1993).

It is usually designed to maintain high biomass concentrations required for fermentation processes. The design of the fermentor and its mode of operation depends on the production organism, the optimal operating conditions required for target product formation, product value, and the scale of production. The performance of any fermenter depends on many factors, but the key physical and chemical parameters that must be controlled are agitation rate, oxygen transfer, pH, and temperature and foam production.

19.1.2.2 Design of Fermentor

Fermentors are constructed with materials like stainless steel that can withstand steam sterilization, vacuum pressure and are non-toxic and non-corrosive. Conventional fermentors are cylindrical vessels surrounded by a jacket and sparger at the bottom through which air is introduced. The bottom portion has a motor to which the agitator shaft is connected. The fermentor has multiple ports for controlling the environment inside the fermentor by regulating the pH, temperature, dissolved oxygen, etc. In recent days, fermentors are usually integrated with computers for efficient process monitoring and data acquisition. The different parts of the fermenter and their uses are given (Table 19.1).

Table 19.1 Parts of fermenter and their uses

Parts	Functions
Impeller/agitator	Stir the media continuously and prevent cells from settling down and distribute oxygen throughout the medium. Impeller speed decreases as the size of the fermenter increases
Sparger/aerator	Introduce sterile oxygen to the media in case of aerobic fermentation process
Baffles/vortex breaker	Disrupt vortex and provide better mixing
Stirrer	Mixes the gas bubbles throughout the liquid culture medium
Top-plate	Cover that is generally made of stainless steel
Drain point	Withdraws cells or medium for the continuous fermentation
Inoculation pipe	Helps to port the inoculum inside the fermentor
Drive motor	Drives the impeller shaft
Impeller shaft	Holds the agitator centrally
Inlet air filter	Filter air before it enters the fermenter
Exhaust air filter	Trap and prevent contaminants from escaping
Rota meter	Measure flow rate of air or liquid
Pressure gauge	Measure pressure inside the fermenter
Temperature probe	Measure and monitor change in temperature of the medium during the process
Cooling jacket	Maintain the temperature of the medium throughout the process by allowing steam or cold water to balance the heat generated during the process
pH probe	Measure and monitor pH of the medium
Dissolved oxygen probe	Measure dissolve oxygen in the fermenter
Level probe	Measure the level of medium
Foam probe	Detect the presence of the foam
Sampling point	To obtain samples during the process
Valves	Regulates and controls the flow of liquids and gases
Peristaltic pump	Pumps acid, base, and antifoam into the medium
Heating pad	Provides heat to the medium
Cold finger	It is a pipe that passes cold water inside a vessel to cool the contents

19.1.2.3 Fermentation Process

It involves the propagation of the microorganism and the production of the desired product. Fermentation process is divided depending on the feeding strategy of the culture and medium as follows.

- Batch fermentation
- Continuous fermentation
- Fed batch fermentation

Batch Fermentation

In this method, the medium and culture are initially fed into the vessel and closed. Later oxygen alone is added to the medium, and pH is adjusted. The fermentation is allowed to run for a predetermined period of time, and the product is harvested at the end.

Continuous Fermentation

This is an open system. It involves the removal of culture medium continuously and replacement of them with a fresh sterile medium in a bioreactor. It includes chemostat and turbidostat bioreactors. It is used for the production of antibiotics, organic solvents, ethanol, and single cell proteins.

Fed Batch System

It is a combination of both batch and continuous systems. In this, additional nutrients are added to the fermentors as the fermentation is in progress. This extends the time of operation, but the products are harvested at the end of the production cycle as in batch fermenter.

19.1.2.4 Types of Fermentors

Continuous Stirred Tank Bioreactor

It consists of a vessel, pipes, valves, pumps, agitator, shaft, impeller, and motor (Rehm et al. 1993). It provides a homogeneous environment for cell growth and proliferation. The main target of desired products includes the cells or primary metabolite. Their mechanic agitation produces shear stress which may harm the cultured cells which can be overcome by changing the shape and diameter of the impeller blade or adding bovine serum albumin or dextran. The height and diameter ratio of this bioreactor is between 3:5. In stirred tank fermentor, rotating stirrer and baffle are found either at the top or the bottom. It mainly uses the batch process of fermentation.

Airlift Bioreactor

In this bioreactor, a baffle or a draft tube is present in the vessel through which air is pumped and create bubble in the medium and helps in mixing the content. It does not have any mechanical stirring. There are internal liquid circulation channels, which enable continuous circulatory motion of the medium. Here, the fermentation occurs at a fixed rate of volume and circulation.

Packed Bed Bioreactor

They are tubular reactors which contains an immobilized biocatalyst and a nutrient broth. Enzyme or microbial cells act as biocatalysts since immobilization enhances the stability of the enzyme and these enzymes can be utilized multiple times. The nutrient broth is allowed to continuously flow over the immobilized biocatalyst which releases the desired product into the fluid at the bottom of the culture vessel. As the substrate and product molecules are carried out both in and out of the carrier matrix during the reaction, there is a problem of external mass transfer resistance (Sen et al. 2017).

Fluidized Bed Bioreactor

It is an immobilized cell reactor which is a combination of stirred tank and packed bed continuous flow reactors. In this, beds of regular molecules are suspended in a flowing liquid stream. An adequate amount of gas is introduced into the medium to form a suitable gas-liquid-solid fluidized bed. This can be used for particles such as immobilized enzymes, immobilized cells, and microbial flocs. Immobilized cell particles are retained in the bed by gravity. It is used for wastewater treatment and hydrogen production (Godia and Sola 1995).

Photobioreactor

These are carried out either by exposing to sunlight or artificial illumination. It consists of many tubes which act as receivers or trappers for solar light which is then transferred to the culture using a centrifugal pump. These are majorly used for algal culture which uses temperature range between 25 °C and 40 °C. It can be placed either vertically or horizontally (Ugwu et al. 2008). During daytime, the microorganisms absorb the light and grow while during night the desired products are formed.

Membrane Bioreactor

It consists of a biological reactor with suspended biomass and solid removal by ultra- and microfiltration membranes. It is used for alcoholic fermentation, solvents, organic acid production, and wastewater treatment. It involves high-quality effluent through the membranes and eliminates the sedimentation and filtration processes. The membrane materials mostly used are polysulfonate, polyamide, and cellulose acetate.

Bubble Column Bioreactors

The bioreactor involves agitation by density-driven fluid motions, and sparger is used to provide air into the continuous liquid phase at the bottom of the reactors. The fluid is mixed intensively at high gas flow rates when convective flows become turbulent. However, back mixing is a major drawback that adversely affects product conversion (Lübbert 2000).

19.1.2.5 Fermentation Medium

Most fermentation requires liquid media, often referred to as broth, although some solid-substrate fermentations are also operated. Fermentation media must satisfy all the nutritional requirements of the microorganism and fulfil the technical objectives of the process. Animal fats and plant oils are also incorporated into some media, often as supplements to the main carbon source. Some fermentation must be supplemented with specific precursors in controlled quantities and in a relatively pure form, notably for secondary metabolite production. Examples: Phenyl acetic acid or phenylacetamide added as side chain precursors in penicillin production. A good medium must have the following characteristics:

- It should be cheap and easily available.
- It should minimize the formation of undesired products and maximize the formation of undesired product.
- It must contain carbon, nitrogen and energy source, micronutrients such as minerals and vitamins required for the industrial production.

19.1.3 Downstream Processing

Various stages that follow the fermentation process in downstream processing involve suitable techniques and methods for recovery, purification, and characterization of the desired fermentation product. Centrifugation, filtration, and chromatography may be applied. These methods vary according to the chemical and physical nature, as well as the desired grade, of the final product. An essential step in downstream processing is quality control, where the product is not only assessed for sterility but also for the absence of genetic materials such as viruses or nucleic acids, and absence of pyrogens, and also of the purity of the final product, such as the desired protein.

The biological products required for industrial purposes can be obtained either from natural sources such as blood, plants, microorganisms, or from recombinant material expressed intracellularly or secreted into culture medium and feedstocks. In both the cases, it becomes essential to purify the product of interest from other biological contaminants. This process wherein the target biological products such as proteins, peptides, DNA, and virus particles are recovered from other contaminants is termed as downstream processing (Bracewell et al. 2009). This downstream processing is carried out using adsorptive chromatographic procedures including high-performance liquid chromatography (HPLC) and size-exclusion chromatography (SEC). The steps involved in the process are determined by the nature of the molecule and the source and how the material is presented for processing.

19.1.3.1 Initial Considerations and Primary Recovery

The biological products are mostly derived using recombinant DNA technology and cell expression systems for which *E. coli* and mammalian cell expression systems are used. In case of *E. coli*, proteins are expressed intracellularly, whereas in mammalian cell expression systems, the proteins are expressed through the secretory pathway and secreted into the surrounding culture media which can be directly recovered from the culture media. But in the case of expression in *E.coli*, the cell must be break open (lysis) in order to release the intracellular recombinant protein.

Centrifugation and Filtration

The first step involved in any downstream processing is centrifugation and filtration. Centrifugation is used either to collect cells where intracellular material needs to be released or to remove cell debris where the biomolecule of interest is in the culture medium. Due to size limitation, batch centrifugation on a large scale is not feasible and therefore, continuous flow centrifugation can be used. In order to clarify cell culture supernatants and cell extracts, both centrifugation and filtration can be used. However, on a large scale, traditional filtering is not possible due to the membrane size that would be required and fouling of the membrane. Cross-flow filtration can be used to reduce both fouling of the membrane and improve the flow rate which has been successfully used to remove soluble host proteins during the recovery of inclusion bodies in *E. coli* cell lysates (Venkiteshwaran et al. 2007).

Cell Lysis

In case of biological products that are derived from *E. coli* where the products are expressed intracellularly, the material must be released out so that it can be recovered and purified. This can be done by enzymatic, chemical, and physical lysis methods. Both chemical and enzymatic approaches can be used to help selectively release target products, but this requires the optimization of each procedure for individual target molecules (Schutte and Kula 1990). Chemical approaches use detergents or high pH alkaline solutions to disrupt and lyse the cell, both of which can adversely affect the molecule of interest if these approaches are not optimized for individual targets. Enzymatic lysis can also be used to give specificity to both the process of lysis and release of the target product (Asenjo and Andrews 1990). Physical approaches include grinding (e.g., in a mortar), high pressure (e.g., in a French press), and osmotic shock. In the case of heat-stable products, heat lysis methods can be employed.

Recovery of Material from Inclusion Bodies

The high-level expression of recombinant protein using *E. coli* expression systems often results in the formation of inclusion bodies which are insoluble aggregates of material that form in vivo (Singh and Panda 2005). When *E. coli* host is unable to fold and process the polypeptide chain(s) of the recombinant protein correctly, it results in inclusion bodies. The protein material found within inclusion bodies has no biological activity, and these inclusion bodies can be easily isolated from *E. coli* at high purities (Singh and Panda 2005). Once the inclusion bodies have been purified, it is often possible to recover the biological activity of the protein of interest. The recovery of protein from inclusion bodies can be low, and the process can be expensive. Further, inclusion bodies are not usually absolutely pure and there may be significant loss of protein during the recovery of material from inclusion bodies. Hence careful attention must be paid to the design of the strategies employed for the recovery of material from inclusion bodies to ensure good recovery and purification of the protein of interest. There are four steps involved in the purification and biological recovery of proteins within inclusion bodies (Singh and Panda 2005) which are as follows:

- Isolation of the inclusion bodies forms the host *E. coli* cells
- Re-solubilization of the protein material within the inclusion bodies

- Refolding of the resolubilized material into a bioactive form
- Purification or polishing of the protein of interest

The most important steps are the solubilization and refolding. Inclusion bodies and therefore the proteins contained within are usually resolubilized using high concentrations (6–8 M) of chaotropic agents such as urea. The resolubilized proteins are then refolded by slowly removing the resolubilizing agent while transferring the process to an appropriate buffer system and concentration in order to reach and retain the correct biological three-dimensional structure and activity.

19.1.3.2 Protein Precipitation

Precipitation of protein from culture medium or following cell lysis is the first step in the purification of a protein which can result in a yield of around 80% with a purification factor in the region of fourfold on average depending on specific proteins. The most commonly used precipitating agent is ammonium sulfate, although other agents such as acetone and polyethylene glycol are also used. Ammonium sulfate is preferred due to its relatively low cost, high solubility, and low toxicity.

19.1.3.3 Chromatography

The major step in downstream process is purification of the target molecule by chromatography. There are four main chromatographic techniques used for the purification of proteins, ion-exchange, affinity, gel filtration, and hydrophobic interaction. Of the four, ion-exchange, affinity and hydrophobic interaction chromatography are absorptive processes whereby binding and elution are undertaken.

Alternatives to Packed Bed Chromatography

Packed bed chromatography is the most widely used mode of chromatography used on both large and small scale. However, due to its high cost and low throughput, some alternative has been evolved which are as follows:

- Expanded bed adsorption
- Aqueous two-phase extraction
- Membrane chromatography and filtration
- Crystallization
- Monolith columns

19.1.3.4 Design of Biomolecules for Downstream Processing

To date, the design of molecules for downstream processing has largely been limited to the use of affinity tags and fusion proteins for purification. These have been particularly adopted at the small laboratory scale but are often not practical, too costly or potentially have legislative issues associated with them, which have prevented such approaches becoming more widely adopted at large scale. However, the design of molecules to aid in more efficient traditional chromatographic approaches, such that fewer steps and greater yields are achieved and/or for aiding alternative approaches such as protein crystallization and aqueous phase partitioning is an attractive proposition. Further, careful design of biomolecules may also be used in the future to prevent or limit aggregation and harmful post-translation modifications occurring (e.g., deamidation) during the expression of the protein of interest in fermenters and while undergoing downstream processing (Cromwell et al. 2006).

19.2 Role of Bioprocess Engineering in Fisheries

Many of the bioactive compounds discovered and isolated from various marine organisms through bioprospecting have led to minimal commercialization due to the limited availability of the compounds for clinical trials or further modification by chemical or biocatalytic means. Hence bioprocess engineering plays a major role in cultivating marine microorganisms to produce high value products. Marine microorganisms have wide distribution and hence may require special cultivation conditions with respect to temperature and pressure. Microorganisms found in the sea usually have a specific requirement for the sodium ion in addition to growing optimally at the water activity of seawater. Some of the widely used media for marine prokaryote culture include marine broth, Luria-Bertani medium and artificial seawater medium. Fermentation protocols are developed based on the temperature and pressure dependency of the marine microorganism for their growth or production (Lang et al. 2005).

Vazquez et al. (2019) developed alternatives to valorize fish discards in order to recover and produce fish mince, oils, gelatins, fish protein hydrolysates (FPHs) with antioxidant and antihypertensive properties, and marine peptones. They studied microbial bioconversion of marine peptones by the growth of *Pediococcus acidilactici* and produced highly digestible FPHs. They also isolated fish oils with a healthy ω-3/ω-6 ratio in the same bioprocess. They concluded that use of peptones derived from FPHs as bacterial media nutrient (source of organic nitrogen) led to reduction of bioproduction costs between two-folds and five-folds in comparison with commercial culture medium.

Oil pollution of water bodies from industrial effluents and occasional oil spills demand effective and low-cost methods for removing them. Wastewater from fish processing industry is also characterized by high salinity and high fat, oil, and grease content. These highly saline and oily wastewater is hard to treat by conventional physicochemical and biological methods. Patel et al. (2022) used a marine thraustochytrid strain that could survive and utilize 120 g/L waste cooking oil (WCO) under highly saline conditions to convert low-quality oil into high-quality microbial lipids rich in docosahexaenoic acid (DHA) and squalene. They further improved the DHA and squalene levels via the co-utilization of hydrophilic and hydrophobic substrates via de novo and ex novo fermentation.

References

Asenjo JA, Andrews BA (1990) Enzymatic cell lysis for product release. Bioprocess Technol 9: 143–175

Bracewell DG, Mumtaz MA, Smales CM (2009) Downstream processing. In: Molecular biology and biotechnology, 5th edn. RSC Publishing, Cambridge, pp 492–512

Cromwell ME, Hilario E, Jacobson F (2006) Protein aggregation and bioprocessing. AAPS J 276: 31–41

Godia F, Sola C (1995) Fluidized-bed bioreactors. Biotechnol Prog 11(5):479–497. https://doi.org/10.1021/bp00035a001

Lang S, Hüners M, Lurtz V (2005) Bioprocess engineering data on the cultivation of marine prokaryotes and fungi. Adv Biochem Engin/Biotechnol 97:29–62. https://doi.org/10.1007/b135822

Lübbert A (2000) Bubble column bioreactors. In: Bioreaction engineering. Springer, Berlin, Heidelberg. https://doi.org/10.1007/978-3-642-59735-0_9

Patel A, Vellosillo ID, Rova U, Matsakas L, Christakopoulos P (2022) A novel bioprocess engineering approach to recycle hydrophilic and hydrophobic waste under high salinity conditions for the production of nutraceutical compounds. Chem Eng J 431:133955

Rehm HJ, Reed G, Puhler A, Stadler P (1993) Biotechnology. In: Bioprocessing, vol 3, 2nd edn. VCH Publishers, Weinheim & New York, p 816

Schutte H, Kula MP (1990) Pilot- and process-scale techniques for cell disruption. Biotechnol Appl Biochem 12:599–620

Sen P, Nath A, Bhattacharjee C (2017) Packed-bed bioreactor and its application in dairy, food, and beverage industry. In: Current developments in biotechnology and bioengineering. Bioprocesses, bioreactors and controls. Elsevier, Amsterdam, pp 235–277

Singh SM, Panda AK (2005) Solubilization and refolding of bacterial inclusion body proteins. J Biosci Bioeng 99:303–310

Ugwu CU, Aoyagi H, Uchiyama H (2008) Photobioreactors for mass cultivation of algae. Bioresour Technol 99(10):4021–4028. https://doi.org/10.1016/j.biortech.2007.01.046

Vazquez JA, Compas AF, Blanco M, Amado IR, Moreno H, Borderias J, Perez-Martin RI (2019) Development of bioprocesses for the integral valorisation of fish discards. Biochem Eng J 144: 198–208

Venkiteshwaran A, Heider P, Matosevic S, Bogsnes A, Staby A, Sharfstein S, Belfort G (2007) Optimized removal of soluble host cell proteins for the recovery of met-human growth hormone inclusion bodies from Escherichia coli cell lysate using crossflow microfiltration. Biotechnol Prog 23:667–672

Bioinformatics: Databases and Formats 20

The term bioinformatics was coined by Paulien Hogeweg in 1979 for the study of informatics processes in biotic systems. Bioinformatics can be defined as "science of solving biological problem using a mathematical and computational approach." Hence it is a scientific discipline that encompasses all the aspects of biological information, viz. acquisition, processing, storage, distribution, analysis, and interpretation, that combines the tools and techniques of mathematics, computer science, and biology with the aim of understanding the biological significance of a variety of data.

Common activities in bioinformatics include mapping and analyzing DNA and protein sequences, aligning different DNA and protein sequences to compare them and creating and viewing 3-D models of protein structures. The primary goal of bioinformatics is to uncover the wealth of biological information hidden in the mass of data and obtains a clear insight into the fundamental biology of organism. Major research efforts in the field include sequence alignment, gene finding, genome assembly, drug design, drug discovery, protein structure alignment, protein structure prediction, prediction of gene expression and protein–protein interactions, genome-wide association studies, and the modeling of evolution.

20.1 Databases

Database is essential for bioinformatics research and applications. Many databases exist, covering various information types like DNA and protein sequences, molecular structures, phenotypes, and biodiversity. It contain empirical data (obtained directly from experiments), predicted data (obtained from analysis), or both. It may be specific to a particular organism, pathway, or molecule of interest. Alternatively, they can incorporate data compiled from multiple other databases. Databases vary in their format, access mechanism, and whether they are public or not.

C. J. Betsy, C. Siva, *Fisheries Biotechnology and Bioinformatics*,
https://doi.org/10.1007/978-981-99-6991-3_20

20.2 Classifications of Databases

- Primary databases: experimental results directly form a database.
- Secondary databases: results of analysis of primary databases.

20.2.1 Primary Database

It is a database consisting of data derived experimentally such as nucleotide sequences, and three-dimensional structures are known as primary database. Primary databases contain information and annotations of DNA and protein sequences, DNA and protein structure, and DNA and protein expression profiles.

20.2.1.1 Types of Primary Database

- Genome database
  - Sequence database
- Structural database
- Protein database
  - Structural database
- Complex database
  - Protein nucleic acid complex database

20.2.2 Secondary Database

It contain results of analysis of primary data including information as sequences patterns of methods, variants, mutation, and evolutionary relationships.

- Protein database
  - Sequence database
- Complex database
  - Protein nucleic acid interaction database

20.2.3 Biological Databases

- Sequence databases
- Structure databases

20.2.3.1 Sequence Databases

- Nucleotide databases
- Protein databases

Nucleotide Databases

- NCBI
- GenBank
- EMBL
- DDBJ

NCBI

The National Centre for Biotechnology information (NCBI) is part of the United States National Library of Medicine (NLM), a branch of the National Institutes of Health. The NCBI is located in Bethesda, Maryland. The NCBI houses genome sequencing data in Genbank and an index of biomedical research articles in PubMed Central and PubMed, as well as other information relevant to biotechnology. All these databases are available online through the Entrez search engine. The NCBI has had responsibility for making the available GenBank DNA sequence database since 1992. GenBank coordinates with individual laboratories and other sequence databases such as those of the European Molecular Biology Laboratory (EMBL) and the DNA Data bank of Japan (DDBJ).

GenBank Sequence Database

The GenBank sequence database is an open access, annotated collection of all publicly available nucleotide sequences and their protein translations. This database is produced at NCBI. GenBank is built by direct submissions from individual laboratories, as well as from bulk submissions from large-scale sequencing centers.

Direct submissions are made to GenBank using Banklt, which is a web-based form, or the stand-alone submission program, Sequin. Upon receipt of a sequence submission, the GenBank staff assigns an accession number to the sequence and performs quality assurance checks. The submissions are then released to the public database, where the entries are retrievable by Entrez or downloadable by FTP. Bulk submissions of expressed sequence tag (EST), sequence-tagged site (STS), genome survey sequences (GSS), and high-throughput genome sequence (HTGS) data are most often submitted by large-scale sequencing centers. The GenBank direct submissions group also processes complete microbial genome sequences.

Hence, Genbank is the NIH genetic sequence database, an annotated collection of all publicly available DNA sequences. It maintains the following databases and tools:

1. Literature Databases, e.g., PUBMED, OMIM
2. Entrey Databases, e.g., Connoted Domain Databases, 3D Domains
3. Nucleotide Databases, e.g., Expressed sequence Tags (EST)
4. Tools for sequence analysis, e.g., BLAST
5. Tools for 3-D structure display and similarity searching, e.g., Cn3D

EMBL

The European Molecular Biology Laboratory (EMBL) is a molecular biology research institution supported by 20 European countries and Australia as associate member state. In 1974, EMBL was created and it is a non-profit organization funded by public research money from its member states. Website: www.ebi.ac.uk

DDBJ

The DNA Data Bank of Japan is located at the National Institute of Genetics of Japan. It shares its data with EMBL and with GenBank. DDBJ has made an effort to collect as much data as possible mainly from Japanese researchers. Search is done by FASTA and BLAST. The analysis is done by ClustalW.

Protein Databases

Protein databases are the most comprehensive source of information on protein. It is necessary to distinguish between universal databases covering protein from all species and specialized data collections storing information about specific families or protein or a group or an organism. Types include,

- Primary Protein Databases like UnitProt/Swiss-Prot
- Secondary Protein Databases like Interpro
- Specialized Protein Databases like GOA, ENZYME
- Structure Databases like PDB

UniProt

It stands for Universal Protein resources. It is a central repository or protein sequences and function, created by joining the information contained in UniProt/SwissProt.

PDB

PDB stands for Protein Data Bank. The RCSB PDB provides a variety of tools and resources for studying the structures of biological macromolecules and their relationships to sequence, function, and disease.

PDB Viewer

It provides an introduction to macromolecular modeling with Deep View, including review of many basic concepts in protein structure. Structure files are used for viewing, and then carry out exercise in manipulating, analyzing, and comparing protein structures.

Swiss-Pdb Viewer

Deep View (formerly called Swiss-Pdb Viewer) is a friendly but powerful molecular graphics program. It is designed for use with computing tools available from the Expert Protein Analysis System, or ExPASy Molecular Biology Server in Geneva, Switzerland.

- It allows us to build models, by giving an amino-acid sequence.
- It can find hydrogen bonds within proteins and between proteins and ligands.
- It allows us to view several proteins simultaneously and superimpose them to compare their structures and sequences.
- It computes electrostatic potentials and molecular surfaces, and carries out energy minimization.

Swiss-Prot

Swiss-Prot is a protein sequence database found out in 1986. It is a manually curated biological database of protein sequences. Swiss-Prot was created in 1986 by Amos Bairoch during his PhD and developed by the Swiss Institute of Bioinformatics and the European Bioinformatics Institute. Swiss-Prot strives to provide reliable protein sequences associated with a high level of annotation (such as the description of the function of a protein, its domains structure, post-translational modifications, variants, etc.), a minimal level of redundancy, and high level of integration with other databases. In 1996, a computer annotated supplement to SWISS-PROT was created, termed TrEMBL.

TrEMBL

It was created in 1996 as a computer annotated supplement to SWISS PROT. The database helps the SWISS PROT format and contains translations of all coding sequences (CDS) in EMBL. It has two main sections:

1. SP-TrEMBL: (SWISS PROT—TrEMBL): it contains the entries that eventually be incorporated into SWISS PROT that have not yet been manually annotated.
2. REM-TrEMBL: it contains sequences that are not destined to be included in SWISS PROT. These include immunoglobulins, T-cell receptors, fragments of fewer than eight amino acids, synthetic sequences, patented sequences, and codon translations.

20.3 Tools for Sequence Analysis

20.3.1 BLAST (Basic Local Alignment Search Tool)

It performs "local" alignment. It finds regions of local similarity between sequences. The program compares nucleotide or protein sequences to sequence databases and calculates the statistical significance of matches. Following are the types of BLAST:

- Blastn: searches a nucleotide database using a nucleotide query.
- Blastp: searches a protein database using a protein query.
- Blastx: searchers a protein database using a translated nucleotide query.
- Tblastx: searches a translated nucleotide database using a translated nucleotide query.

20.3.2 CLUSTALW

Clustalw is a multiple sequence alignment program for proteins or nucleotides which is available at <ebi.ac.uk/clustalw>. Here the multiple sequence alignment means an extension of pairwise alignment to incorporate more than two sequences regions across a group of sequences hypothesized to be evolutionarily related. It calculates the best match for the selected sequences so that differences, similarities, and identities can be seen.

- Evolutionary relationship can be seen using cladograms and phylograms. Phylogram is a branching diagram (tree) assumed to be an estimate of a phylogeny, and branch lengths are proportional to the amount of inferred evolutionary change. A cladogram is a branching diagram (tree) assumed to be a phylogeny where the branches are of equal length, thus cladograms show common ancestry, but do not indicate the amount of evolutionary "time" separating taxa.
- It can align either nucleotide or protein sequences.
- The program accepts sequences in the formats like-NBRF/PIR, EMBL/UniProt, Pearson (Fasta), GDE, ALN/ClustalW. The sequences can either be pasted into the web form or uploaded to the web form in a file.

20.3.3 FASTA

Fasta is a protein similarity search. It provides sequence similarity searching of query sequence against nucleotide and protein databases using fasta programs. It can be used for fast protein comparison or fast nucleotide comparison. This format contains a one-line header followed by lines of sequence data. The sequence in fasta formatted files is preceded by a line starting with a ">" symbol. The first word on this line is the name of the sequence. The rest of the line is a description of the sequence. The remaining lines contain the sequence itself. Fasta files containing multiple sequences are just the same, with one sequence listed right after another. This format is accepted for many multiple sequence alignment programs.

21 Sequence Retrieval, Analysis, and Manipulation

In the early 1960s, when computers capable of analyzing sequences were unavailable, known protein sequences accumulated slowly. In this pre-computer era, sequences were assembled, analyzed, and compared by (manually) writing them on pieces of paper, taping them side by side on laboratory walls, and/or moving them around for optimal alignment which is now called pattern matching. When computers became available, the first computational biologists started to enter these manual algorithms into the memory banks which was brand new. Most methods had to be invented from scratch, and in the process, a new area of research, the analysis of protein sequences using computers was generated. This was the genesis of bioinformatics.

Over the years, the amount of DNA and protein sequence data generated in biology has grown enormously and even so, continues to grow exponentially. As these sequences are discovered, they are typically placed in public online databases where researchers can readily access them. Expectedly, the available public DNA and protein sequence databases are huge. In bioinformatics therefore, sequence retrieval and analysis are very important.

The central dogma of molecular biology shows the expression of genes from DNA to RNA and then to protein. It involves the process of transfer of genetic information from DNA to DNA (replication), DNA to RNA (transcription), and RNA to protein (translation). These biomolecules contain sequences of nucleotides (nucleobases) which need to be studied and analyzed in order to decipher the precise order of the nucleobases in them in a process of sequence retrieval and analysis. When these sequences are retrieved and analyzed, they form a viable tool in the study of heredity and variation, diagnosis and treatment of diseases, and production of drugs.

C. J. Betsy, C. Siva, *Fisheries Biotechnology and Bioinformatics*,
https://doi.org/10.1007/978-981-99-6991-3_21

21.1 Sequence Retrieval System

This is the operation of accessing the precise order of gene in a DNA, RNA, and protein. Sequence information comes from many sources. Many methods have been employed in determining the genomic sequence including the Sangers, manual and automated methods but the recent knowledge of bioinformatics proffered an efficient and effective way of accessing the sequence using computer. When the name of a gene or its ID number is given, it is possible to find and retrieve its DNA sequence. Sequence of genes are given an accession number as identification for database processing. Each sequence has its own unique accession number, but there may be some sequences that have more than one accession number. The most consistent source of sequence data comes from sequencing centers. The foundation of online computer database for storing and distributing sequence data has made bioinformatics an invaluable tool for sequence retrieval.

21.1.1 Types of Sequence Retrieval Databases

- Global nucleotide sequence storage
- Genome centered database
- Protein database

21.1.1.1 Global Nucleotide Sequence Storage

- National Centre for Biotechnology Information (NCBI)—(www.ncbi.nlm.nih.gov/)
- European Molecular Biology Laboratory (EMBL)—(www.ebi.ac.uk/embl/)
- DNA Database of Japan (DDBJ)—(www.ddbj.nig.ac.jp/)

They collect all publicly available DNA, RNA and protein sequence data and make it available for free. Due to their daily exchange of data, these all databases contain essentially the same data.

21.1.1.2 Genome Centered Database

- NCBI genomes: Entrez Life Sciences Search Engine (US National Institutes of Health)—www.ncbi.nlm.nih.gov/sites/gquery
- Ensemble genome browser (European Bioinformatics Institute)—www.ensemble.org
- UCSC genome bioinformatics site (University of California at Santa Cruz)—www.genome.ucsc.edu

21.1.1.3 Protein Database

- Swiss-Prot
- TrEMBL
- PDB

21.1.2 Sequence Retrieval in NCBI

The National Centre for Biotechnology Information (NCBI) is U.S. government funded national resource for molecular biology information. NCBI database contains essentially the same data as in the EMBL/DDBJ databases. Sequences in the NCBI Sequence Database (or EMBL/DDBJ) are identified by an accession number. This is a unique number that is only associated with one sequence. For example, the accession number NC_001477 is for the DEN-1 Dengue virus genome sequence. The accession number identifies the sequence. It is reported in scientific papers describing that sequence. For each sequence, the NCBI database (or EMBL/DDBJ databases) also stores some additional annotation data, such as the name of the species it comes from, references to publications describing that sequence, etc. Some of this annotation data was added by the person who sequenced the sequence and submitted it to the NCBI database, while some may have been added later by a human curator working for NCBI.

The NCBI database contains several sub-databases, the most important of which are the "NCBI Nucleotide database" which contains DNA and RNA sequences, the "NCBI Protein database" which contains protein sequences, "EST" contains ESTs (expressed sequence tags), which are short sequences derived from mRNAs, the "NCBI Genome database" which contains DNA sequences for whole genomes. PubMed contains data on scientific publications.

The sequence retrieval system is a homogeneous interface to over 80 biological databases that had been developed at the European Bioinformatics Institute (EBI) at Hinxton, UK. It includes databases of sequences, metabolic pathways, transcription factors, application results (like BLAST, SSEARCH, FASTA), protein 3-D structures, genomes, mappings, mutations, and locus specific mutations. The foundation of online computer database for storing and distributing sequence data has made bioinformatics an invaluable tool for sequence retrieval.

21.2 Sequence Analysis

Sequence analysis refers to the process of subjecting DNA, RNA, or peptide sequence to any analytical procedure aimed at understanding its features, function, structure, or evolution. A number of methodologies are available for carrying out sequence analysis. They include sequence alignment, searches against databases, and others. Sequence analysis is especially important because the rate of addition of new gene and protein sequences to the databases has increased exponentially. However, by comparing these new sequences to those with known functions, the understanding of the biology of the organism from which the new sequence came can be achieved. Sequence analysis, therefore, can be used to assign functions to genes and proteins by the study of the similarities between the compared sequences. Many tools and techniques are available lately to provide sequence comparisons (sequence alignment) and analyze the alignment product to understand its biology.

21.2.1 Output of Sequence Analysis

1. The comparison of sequences in order to find similarity, to infer if they are related (homologous)
2. Identification of intrinsic features of the sequence such as active sites, post translational modification sites, gene structures, reading frames, distributions of introns and exons, and regulatory elements
3. Identification of sequence differences and variations such as point mutations and single nucleotide polymorphism (SNP) in order to get the genetic marker
4. Revealing the evolution and genetic diversity of sequences and organisms
5. Identification of molecular structure from sequence alone

21.2.2 Parameters for Which Sequence Analysis Is Applied

- Sequence alignment
- Profile comparison
- Sequence assembly
- Gene prediction
- Protein structure prediction

21.2.2.1 Sequence Alignment

There are millions of protein and nucleotide sequences known. These sequences fall into many groups of related sequences known as protein families or gene families. Relationships between these sequences are usually discovered by aligning them together and assigning this alignment a score. There are two main types of sequence alignment. Pairwise sequence alignment only compares two sequences at a time, and multiple sequence alignment compares many sequences. Two important algorithms for aligning pairs of sequences are the Needleman–Wunsch algorithm and the Smith–Waterman algorithm. Popular tools for sequence alignment include the following:

- Pairwise alignment—BLAST, Dot plots
- Multiple alignment—ClustalW, PROBCONS, MUSCLE, MAFFT, and T-Coffee

Alignment Methods

Alignment is done to compare a sequence with sequences that have already been reported and stored in a database. It can be either global and local. Local alignments reveal regions that are highly similar, but do not necessarily provide a comparison across the entire two sequences. The global approach compares one whole sequence with other entire sequences.

Dot Plot

Dot plot visualizes the similarity between two nucleic acids or protein sequences. It can be defined as a graphical method that allows the comparison of two biological sequences so as to identify regions of close similarity between them. It is a matrix that is used to visually detect alignments. It involves constructing a matrix with one of the sequences to be compared running horizontally across the bottom, and the other running vertically along the left-hand side. It can compare two sequences by organizing one sequence on the *x*-axis and another on the *y*-axis of a plot.

It uses simplest scoring system which distinguishes only between identical (dots) and non-identical (blank) residues. Identical sequences will obviously have a diagonal line in the center of the matrix. Insertions and deletions between sequences give rise to disruptions in this diagonal. Regions of local similarity or repetitive sequences give rise to further diagonal matches in addition to the central diagonal. However, it is helpful only in very short sequences.

21.2.2.2 Profile Comparison

In 1987, Michael Gribskov, Andrew McLachlan, and David Eisenberg introduced the method of profile comparison for identifying distant similarities between proteins. It uses a multiple sequence alignment to encode a profile which contains information about the conservation level of each residue. These profiles can then be used to search collections of sequences to find sequences that are related. Profiles are also known as Position Specific Scoring Matrices (PSSMs). In 1993, a probabilistic interpretation of profiles was introduced by David Haussler and colleagues using Hidden Markov Models. These models are known as profile-HMMs.

21.2.2.3 Sequence Assembly

Sequence assembly refers to the reconstruction of a DNA sequence by aligning and merging small DNA fragments. It is an integral part of modern DNA sequencing. Since presently available DNA sequencing technologies cannot read long sequences, large pieces of DNA (such as genomes) are often sequenced by cutting the DNA into small pieces, reading the small fragments, and reconstituting the original DNA by merging the information on various fragments.

21.2.2.4 Gene Prediction

Gene prediction or gene finding refers to the process of identifying the regions of genomic DNA that encode genes. This includes protein-coding genes as well as RNA genes, but may also include the prediction of other functional elements such as regulatory regions. Gene finding is one of the first and most important steps in understanding the genome of a species once it has been sequenced. In general, the prediction of bacterial genes is significantly simpler and more accurate than the prediction of genes in eukaryotic species that usually have complex intron/exon patterns. Identifying genes in long sequences remains a problem, especially when the number of genes is unknown. HMM can be part of the solution. Traditional sequencing analysis focused on the statistical parameters of the nucleotide sequence itself. Another method is to identify homologous sequences based on the other known gene sequences.

21.2.2.5 Protein Structure Prediction

The 3D structures of molecules are of great importance to their functions in nature. Since structural prediction of large molecules at an atomic level is a largely intractable problem, some biologists introduced ways to predict 3D structure at a primary sequence level. This includes the biochemical or statistical analysis of amino acid residues in local regions and structural inference from homologs (or other potentially related proteins) with known 3D structures.

21.3 Sequence Manipulation Suite

The sequence manipulation suite is a collection of web-based programs for analyzing and formatting DNA and protein sequences. The output of each program is a set of HTML commands, which is rendered by your web browser as a standard web page. You can print and save the results, and you can edit them using an HTML editor or a text editor. It can be found at https://bioinformatics.org/sms/

Following are the standard manipulations that can be carried out in the available sequences.

21.3.1 DNA Entry

- Filter DNA
- GenBank Feature Extractor
- GenBank to FASTA
- Random DNA Sequence

21.3.2 DNA Manipulation

- Reverse complement
- Shuffle DNA

21.3.3 DNA Figures

- Group DNA
- Primer show
- Rest and Trans Map
- Show Translation

21.3.4 DNA Analysis

- Codon Plot
- Codon Usage
- CpG Islands
- DNA Pattern Find
- DNA Stats
- ORF Finder
- Positional Base Frequencies
- Restriction Summary
- Simple Plot
- TestCode
- Translate

21.3.5 Protein Entry

- Filter Protein
- GenBank Trans Extractor
- Random Protein Sequence

21.3.6 Protein Manipulation

- One To Three
- Shuffle Protein
- Three To One

21.3.7 Protein Figures

- COMBOSA
- Group Protein
- Multiple Align Show

21.3.8 Protein Analysis

- Ident and Sim
- Protein Stats
- Protein Pattern Find
- Protein Molecular Weight
- Multi Rev Trans
- Reverse Translate

Codon Plot accepts a DNA sequence and generates a graphical plot consisting of a horizontal bar for each codon. The length of the bar is proportional to the frequency of the codon in the codon frequency table you enter. Codon frequency tables for numerous organisms are available online (http://www.kazusa.or.jp/codon/). Use Codon Plot to find portions of DNA sequence that may be poorly expressed, or to view a graphic representation of a codon usage table (by using a DNA sequence consisting of one of each codon type).

Codon Usage accepts a DNA sequence and returns the number and frequency of each codon type. Since the program also compares the frequencies of codons that code for the same amino acid (synonymous codons), you can use it to assess whether a sequence shows a preference for particular synonymous codons.

DNA/Protein Pattern Find accepts a sequence along with a set of search patterns and returns the number and positions of sites that match the patterns. The search patterns can contain "wild cards," which allow you to detect a variety of similar sequences using a single pattern.

DNA/Protein Stats returns the number of occurrences of each residue in the sequence you enter. Percentage totals are also given for each residue, and for certain groups of residues, allowing you to quickly compare the results obtained for different sequences.

Filter DNA/Protein removes non-protein or non-DNA characters from text. Use this program when you want to remove digits and blank spaces from a sequence to make it suitable for other applications. All programs in the sequence manipulation suite automatically filter the sequences you enter before proceeding.

GenBank® Feature Extractor accepts a GenBank file as input and reads the sequence feature information described in the feature table, according to the rules outlined in the GenBank release notes (*ftp://ncbi.nlm.nih.gov/genbank/gbrel.txt*). The program concatenates the relevant sequence segments and returns each sequence feature in FASTA format. GenBank Feature Extractor is particularly helpful when you want to derive the sequence of a cDNA from a genomic sequence that contains many introns.

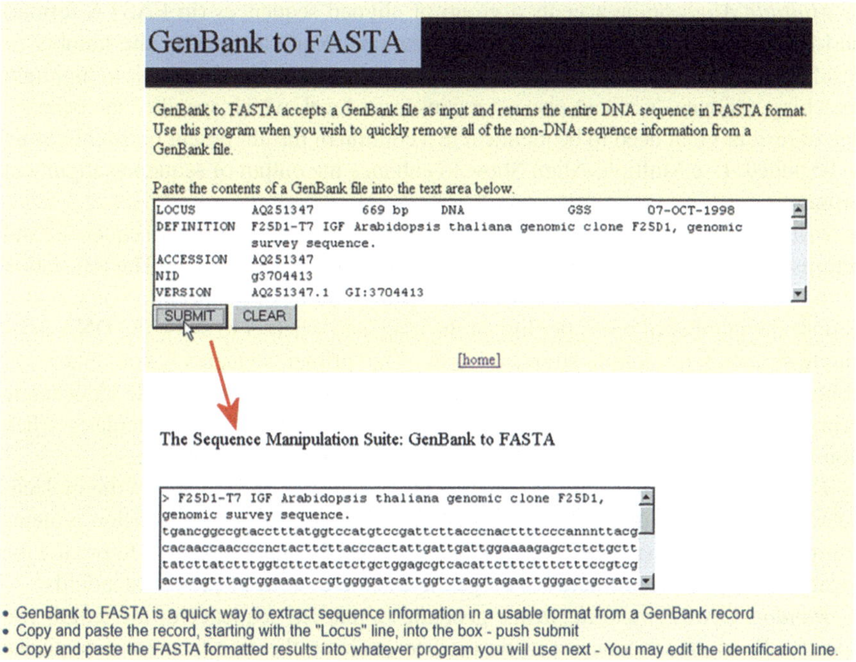

GenBank to FASTA accepts a GenBank file as input and returns the entire DNA sequence in FASTA format. Use this program when you want to quickly remove all of the non-DNA sequence information from a GenBank file.

GenBank Trans Extractor accepts a GenBank file as input and returns each of the protein translations described in the file in FASTA format. GenBank Trans Extractor should be used when you are more interested in the predicted protein translations of a DNA sequence than the DNA sequence itself.

Group DNA/Protein adjusts the spacing of DNA or protein sequences and adds numbering. You can specify the group size (the number of residues per group) and the number of residues per line. The output of this program can serve as a convenient reference for use with some of the other programs in the Sequence Manipulation Suite (such as DNA/Protein Pattern Find) since the numbering and spacing allow you to quickly locate specific residues.

Ident and Sim accepts a pair of aligned sequences (in FASTA format) and calculates their identity and similarity. Identity and similarity values are often used to assess whether or not two sequences share a common ancestor or function. Since many alignment programs do not calculate these values, you may find Ident and Sim quite useful when making pairwise comparisons.

ORF Finder searches for open reading frames (ORFs) in the DNA sequence you enter. The program returns the range of each ORF along with its protein translation. Use ORF Finder to search newly sequenced DNA for potential protein-encoding segments.

Multiple Align Show accepts a group of aligned sequences (in FASTA format) and formats the alignment to your specifications. You can specify the number of residues per line, and whether to use colored text or colored backgrounds to highlight matching residues. You can also set a consensus level, which specifies the percentage of residues that need to be identical in a column of the alignment for highlighting to be added. Use Multiple Align Show to enhance the output of sequence alignment programs.

Primer Show accepts a DNA sequence along with a set of primer sequences and returns a textual map showing the annealing positions of the primers. The translation of the DNA sequence can be shown in the reading frames you specify. You can also choose the number of bases per line of the map, and whether to show the DNA in its single-stranded or double-stranded form. The primer sequences you enter can contain "wild cards," a feature that allows Primer Show to handle degenerate primers. Use this program to produce a useful reference figure, particularly when you have designed a large number of primers for a particular template.

Protein Molecular Weight accepts a protein sequence and calculates the molecular weight. You can append copies of commonly used epitopes and fusion proteins using the supplied list. Use Protein Molecular Weight when you want to predict the location of a protein of interest on a gel in relation to a set of protein standards.

Random DNA/Protein Sequence generates a random sequence of the length you specify. Random sequences can be used to evaluate the significance of sequence analysis results.

Rest and Trans Map accepts a DNA sequence and returns a textual map showing the positions of restriction endonuclease cut sites. The translation of the DNA sequence is also given, in the reading frame you specify. You can choose the number of bases per line of the map, and whether to show the DNA in its single-stranded or double-stranded form. Use the output of this program as a reference when planning cloning strategies, particularly when you need to consider the reading frames of the molecules you are ligating.

Restriction Summary accepts a DNA sequence and returns the number and positions of restriction endonuclease cut sites. Use this program when you want to quickly determine whether or not an enzyme cuts a particular segment of DNA, and to produce a table to complement the output of Rest and Trans Map.

Reverse Complement converts a DNA sequence into its reverse, complement, or reverse complement counterpart. You may want to work with the reverse complement of a sequence if it contains an ORF on the reverse strand.

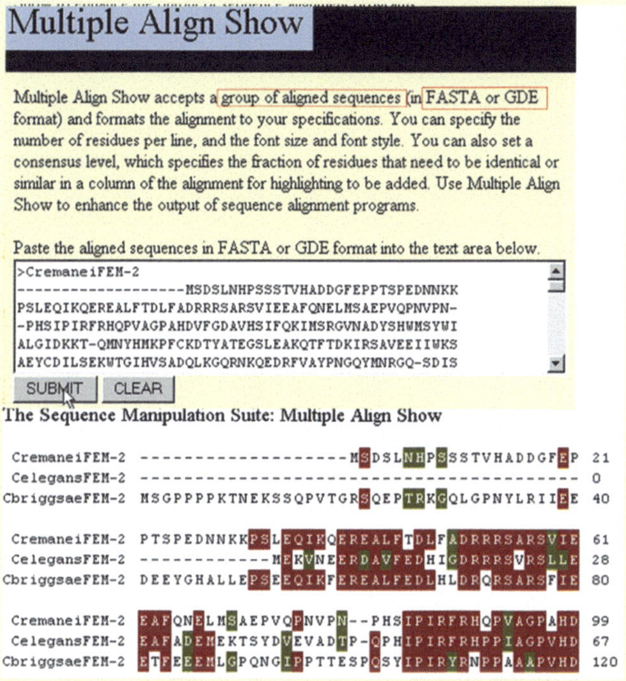

• Multiple Align Show takes aligned sequences, in FASTA or GDE (a ClustalW choice) format and allows you to choose various options

Reverse Translate accepts a protein sequence and uses codon usage table to generate a graph that can be used to find regions of minimal degeneracy at the nucleotide level. Use Reverse Translate to design PCR primers for amplifying genes that encode similar proteins.

Shuffle DNA/Protein randomly shuffles a sequence. Shuffled sequences can be used to evaluate the significance of sequence analysis results, particularly when sequence composition is an important consideration.

Simple Plot calculates DNA sequence composition using a sliding window. The results are returned as a set of x and y values. The x-value is the position of the first base in the window of bases used in the calculation, and the y-value is the result of the calculation. Use Simple Plot when you are looking for segments of a DNA sequence with a particular base composition, rather than a specific sequence of bases.

Test Code accepts a DNA sequence and calculates the Test Code value. The reading frame of the DNA sequence does not need to be considered when performing this analysis. Use Test Code to predict whether or not a sequence encodes a protein.

22 Bioinformatics Tools and Techniques

22.1 Primer Design

A primer is a short synthetic oligonucleotide which is used in many molecular techniques from PCR to DNA sequencing. These primers are designed to have a sequence which is the reverse complement of a region of template or target DNA to which we wish the primer to anneal.

22.1.1 Before Starting

- Check literature and databases for existing primers
- Choose a target sequence
- Design primers (and probes)
- Check primer specificity
- Validate primers

22.1.2 Characteristics of Good Primer

- Design primers that have a GC content of 40–60%
- A melting temperature (T_m) in the range of 50–65 °C (52–58 °C)
- Absence of dimerization and hairpin capability
- The presence of G or C bases within the last five bases from the 3′ end of primers (GC clamp) to enhance annealing of the end which will be extended due to the stronger bonding of G and C bases
- More than 3 G's or C's should be avoided in the last 5 bases at the 3′ end of the primer—mispriming 5′-CCGATATGCCAGCTA*TCTGT*-3′

C. J. Betsy, C. Siva, *Fisheries Biotechnology and Bioinformatics*,
https://doi.org/10.1007/978-981-99-6991-3_22

22.1.2.1 Uniqueness

There shall be one and only one target site in the template DNA where the primer binds, which means the primer sequence shall be unique in the template DNA, avoiding the possibility of mis-hybridization to a similar sequence nearby. There shall be no annealing site in possible contaminant sources such as human, rat, and mouse (BLAST search against corresponding genome).

22.1.2.2 Length

- Primer length has effects on uniqueness and melting/annealing temperature.
  - the longer the primer, the more chance that it is unique
  - the longer the primer, the higher melting/annealing temperature—specificity
- The length of primer has to be at least 15 bases to ensure uniqueness.
- Usually, we pick primers of 17–28 bases long.
- This range varies based on if you can find unique primers with appropriate annealing temperature within this range.
- Above 30 bases leads to risk of mispairing, primer dimers, and hairpins.

22.1.2.3 Base Composition

It affects hybridization specificity and melting/annealing temperature. Random base composition is usually preferred. Avoid long (A + T) and (G + C) rich region if possible. A repeat is a dinucleotide occurring many times consecutively and should be avoided because they can misprime. A maximum number of dinucleotide repeats acceptable in an oligo is four dinucleotides.

Template DNA 5′..TCGATATATATGCATG... GATGCCGGCGCGCTGTACACAA.0.3′.

Usually, average (G + C) content around 40–60% will give us the right melting/annealing temperature for ordinary PCR reactions and will give appropriate hybridization stability. Primers with long runs of a single base should generally be avoided as they can misprime. For example, AGCGGGGGATGGGG has runs of base "G" of value 5 and 4. Avoiding repeats of more than 3 to 4 bases is accepted.

22.1.2.4 Melting Temperature

This is the temperature at which 50% of the primer molecules are bound to their corresponding target sequence. T_m is characteristic of the DNA/base composition. Higher G + C content DNA has a higher T_m due to more hydrogen bonds.

$$T_m = 64.9 + 41^*(yG + zC - 16.4)/(wA + xT + yG + zC)$$

22.1.2.5 Annealing Temperature

T_{anneal} is the temperature at which primers anneal to the template DNA. It can be calculated from T_m.

$$T_{\text{anneal}} = T_{\text{m\_primer}} - 4\,^{\circ}\text{C}$$

Too high T_{anneal} will produce insufficient primer-template hybridization resulting in low PCR product yield. Too low T_{anneal} may possibly lead to non-specific products caused by a high number of base pair mismatches. Mismatch tolerance is found to have the strongest influence on PCR specificity. The difference in T_{anneal} between forward and reverse primer allowed is maximum 3 °C. The closer their T_{anneal} are, the better. If primers can anneal to themselves or anneal to each other (primer dimer) rather than anneal to the template, the PCR efficiency will be decreased dramatically.

22.1.3 Good Primer

- Uniqueness: ensure correct priming site
- Length: 17–28 bases. This range varies.
- Base composition: average (G + C) content around 40–60%; avoid long (A + T) and (G + C) rich region if possible.
- Optimize base pairing: G or C in the 3′ end but not too many—to minimize false priming.
- Melting T_m between 52 °C and 65 °C are preferred.
- Assure that F/R primers have annealing T within 2–3 °C of each other.
- Minimize internal secondary structure: hairpins and dimers shall be avoided.

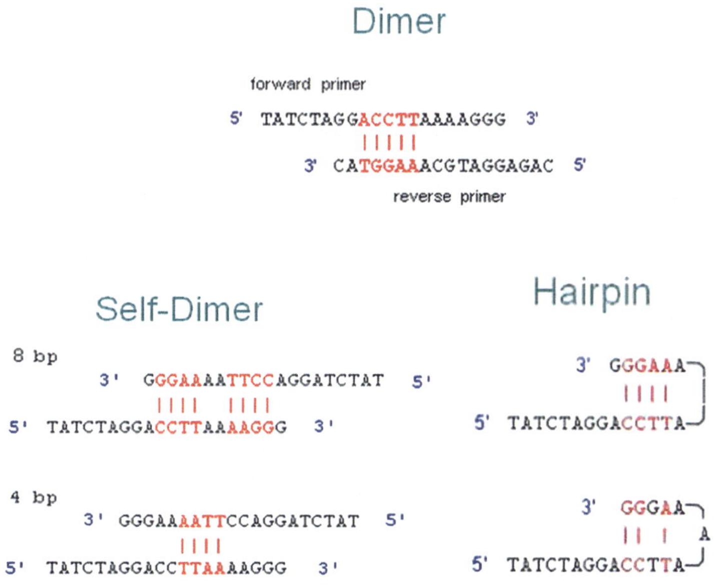

22.2 Restriction Mapping

Restriction mapping is a method used to map an unknown segment of DNA by breaking it into pieces and then identifying the locations of the breakpoints. Restriction maps are used as a reference to engineer plasmids or other relatively short pieces of DNA, and sometimes for longer genomic DNA. This method relies upon the use of proteins called restriction enzymes, which can cut, or digest, DNA molecules at short, specific sequences called restriction sites. After a DNA segment has been digested using a restriction enzyme, the resulting fragments can be examined using a laboratory method called gel electrophoresis, which is used to separate pieces of DNA according to their size.

One common method for constructing a restriction map involves digesting the unknown DNA sample in three ways. Here, two portions of the DNA sample are individually digested with different restriction enzymes, and a third portion of the DNA sample is double digested with both restriction enzymes at the same time. Next, each digestion sample is separated using gel electrophoresis, and the sizes of the DNA fragments are recorded. The total length of the fragments in each digestion will be equal. However, because the length of each individual DNA fragment depends upon the positions of its restriction sites, each restriction site can be mapped according to the lengths of the fragments. The information from the double digestion is particularly useful for correctly mapping the sites. The final drawing of the DNA segment that shows the positions of the restriction sites is called a restriction map.

One approach in constructing a restriction map of a DNA molecule is to sequence the whole molecule and to run the sequence through a computer program that will find the recognition sites that are present for every restriction enzyme known.

22.2.1 Method

The experimental procedure first requires an aliquot of purified plasmid DNA for each digest to be run. Digestion is then performed with each enzyme(s) chosen. The resulting samples are subsequently run on an electrophoresis gel, typically on agarose gel. The first step following the completion of electrophoresis is to add up the sizes of the fragments in each lane. The sum of the individual fragments should equal the size of the original fragment, and each digest's fragments should also sum up to be the same size as each other. If fragment sizes do not properly add up, there are two likely problems. In one case, some of the smaller fragments may have run off the end of the gel. This frequently occurs if the gel is run too long. A second possible source of error is that the gel was not dense enough and therefore was unable to resolve fragments close in size. This leads to a lack of separation of fragments which were close in size. If all of the digests produce fragments that add up one may infer the position of the REN (restriction endonuclease) sites by placing them in spots on the original DNA fragment that would satisfy the fragment sizes produced by all three digests.

22.2.2 Example

For example, the most common application of restriction mapping is determining the orientation of a cloned insert. This method requires that restriction maps of the cloning vector and the insert are already available. If you know of a restriction site placed toward one end of the insert, you can determine the orientation by observing the size of the fragments in the gel. Often the orientation of inserts is important and this technique is used to screen for the correct orientation. In this example, the orientation of an insert cloned with EcoRI will be found.

22.2.2.1 Digests

- 1: EcoRI
- 2: HindIII
- 3: EcoRI + HindIII

22.2.2.2 Resultant Fragments: Approximate Sizes

- 1: 3 kb, 5 kb
- 2: 2 kb, 6 kb
- 3: 2 kb, 1 kb, 5 kb

22.2.2.3 Hypothetical Multiple Cloning Site of Vector

5′-----HindIII-EcoRI----3′.

Discussion

The EcoRI digest excises the insert yielding fragments of 3 and 5 kb. These are the sizes of the insert and vector backbone, respectively. This is expected since the sizes of both insert and vector are known beforehand. The presence of the insert is confirmed.

There is a known HindIII site off-center in the 3 kb insert. It is 2 kb away from one end (end A), and 1 kb away from the other end (end B). The HindIII digest of your clone yields fragments of 2 and 6 kb. The 2 kb fragment is exclusively the insert sequence, and the 6 kb fragment is 1 kb of insert sequence attached to 5 kb of vector sequence. This means that the insert was cloned in an A to B orientation as opposed to B to A which would yield fragments of 7 kb and 1 kb.

Resultant Map

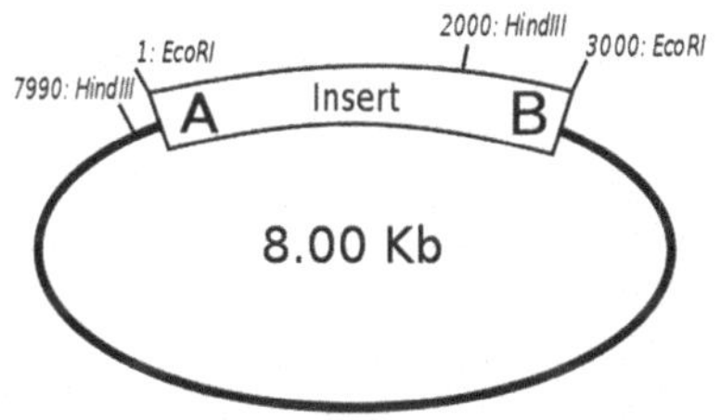

22.3 Orf Finder

22.3.1 Reading Frame

A reading frame is a sequence of nucleotide triplets that are read as codons specifying amino acids; a single strand of DNA sequence has three possible reading frames. Once a gene has been sequenced it is important to determine the correct open reading frame (ORF). Every region of DNA has six possible reading frames, three on forward strand and three on complementary strand. The reading frame that is used determines which amino acids will be encoded by a gene.

22.3.2 Open Reading Frames

ORFs are parts of a reading frame that contain no stop codons. The region of the nucleotide sequences from the start codon (ATG) to the stop codon is called the open reading frame. Gene finding in organism especially prokaryotes starts form searching for an open reading frame (ORF). An ORF is a sequence of DNA that starts with start codon "ATG" (not always) and ends with any of the three termination codons (TAA, TAG, TGA).

While eukaryotic gene finding is altogether a different task as the eukaryotic genes are not continuous and interrupted by intervening noncoding sequences called "introns." Moreover, organization of genetic information in eukaryotes and prokaryotes is different.

An ORF is the part of a reading frame that has the ability to be translated. An ORF has a chain of codons including start and stop codons. The starting point of translation is indicated by ATG codon (AUG in RNA) within the ORF. The transcription termination site is located after the ORF, beyond the translation stop codon. If transcription was to cease before the stop codon, an incomplete protein would be made during translation.

In gene finding, the start-stop definition of an ORF therefore only applies to spliced mRNAs, not genomic DNA, since introns may contain stop codons and/or cause shifts between reading frames. An alternative definition says that an ORF is a sequence that has a length divisible by three and is bounded by stop codons. Such an ORF corresponds to parts of a gene rather than the complete gene.

22.3.2.1 Biological Significance

One common use of ORFs is as one piece of evidence to assist in gene prediction. Long ORFs are often used, along with other evidence, to initially identify candidate protein-coding regions or functional RNA-coding regions in a DNA sequence. The presence of an ORF does not necessarily mean that the region is always translated. For example, in a randomly generated DNA sequence with an equal percentage of each nucleotide, a stop codon would be expected once every 21 codons. A simple gene prediction algorithm for prokaryotes might look for a start codon followed by an ORF that is long enough to encode a typical protein, where the codon usage of that region matches the frequency characteristic for the given organism's coding regions. Therefore, some authors say that an ORF should have a minimal length, e.g., 100 codons or 150 codons. The presence of a long ORF is not a conclusive evidence for the presence of a gene.

22.3.3 ORF Finding Tools

22.3.3.1 ORF Finder

ORF Finder searches for open reading frames (ORFs) in the DNA sequence you enter. The program returns the range of each ORF along with its protein translation. Use ORF Finder to search newly sequenced DNA for potential protein-encoding segments. An ORF is a portion of a DNA molecule that, when translated into amino acids, contains no stop codons.

The ORF Finder is a graphical analysis tool which finds all open reading frames of a selectable minimum size in a user's sequence or in a sequence already in the database. This tool identifies all ORF using the standard or alternative genetic codes. The deduced amino acid sequence can be saved in various formats and searched against the sequence database using the BLAST server. The ORF Finder should be helpful in preparing complete and accurate sequence submissions.

22.3.3.2 ORF Investigator

ORF Investigator is a program which not only gives information about the coding and noncoding sequences but also can perform pairwise global alignment of different gene/DNA regions sequences. The tool efficiently finds the ORFs for corresponding amino acid sequences and converts them into their single letter amino acid code, and provides their locations in the sequence. The pairwise global alignment between the sequences makes it convenient to detect the different mutations, including single nucleotide polymorphism. Needleman–Wunsch algorithms are used for the gene alignment. The ORF Investigator is written in the portable Perl programming language and is therefore available to users of all common operating systems.

22.3.3.3 ORF Predictor

ORF Predictor is a web server designed for identifying protein-coding regions in expressed sequence tag (EST)-derived sequences. For query sequences with a hit in BLASTX, the program predicts the coding regions based on the translation reading frames identified in BLASTX alignments, otherwise, it predicts the most probable coding region based on the intrinsic signals of the query sequences. The output is the predicted peptide sequences in the FASTA format, and a definition line that includes the query ID, the translation reading frame, and the nucleotide positions where the coding region begins and ends. ORF Predictor facilitates the annotation of EST-derived sequences, particularly, for large-scale EST projects.

ORF Predictor uses a combination of the two different ORF definitions mentioned above. It searches stretches starting with a start codon and ending at a stop codon. As an additional criterion, it searches for a stop codon in the 5′ untranslated region (UTR).

22.3.3.4 ORFik

ORFik is a R-package in bioconductor for finding open reading frames and using next generation sequencing technologies for justification of ORFs.

22.4 EMBOSS

European Molecular Biology Open Software Suite (EMBOSS) is an open-source software analysis package developed for the needs of the molecular biology and bioinformatics user community. The software automatically copes with data in a variety of formats and even allows transparent retrieval of sequence data from the web. It integrates a range of currently available packages and tools for sequence analysis into a seamless whole.

The core EMBOSS groups are collaborating with many other groups to develop the new applications that the users need. This was done from the beginning with EMBnet, the European Molecular Biology Network. The EMBOSS package contains a variety of applications for sequence alignment, rapid database searching with sequence patterns, protein motif identification (including domain analysis), and much more.

22.4.1 Advantages of EMBOSS

- A properly constructed toolkit for creating robust bioinformatics applications or workflows.
- A comprehensive set of sequence analysis programs.
- All sequence and many alignment and structural formats are handled.
- Extensive programming library for common sequence analysis tasks.
- Additional programming libraries for many other areas including string handling, pattern-matching, list processing, and database indexing.
- It is free-of-charge.
- It is an open-source project.
- It runs on practically every UNIX, plus MS Windows, and MacOS.
- Each application has the same style of interface.
- The consistent user interface facilitates GUI designers and developers.
- It integrates other popular publicly available packages.
- It is free of arbitrary size limits: there are no limits on the amount of data that can be processed. For the programmer, memory management for objects such as sequences and arrays is simplified.

22.4.2 Applications of EMBOSS

- Sequence alignment
- Rapid database searching with sequence patterns
- Protein motif identification, including domain analysis
- Nucleotide sequence pattern analysis. For e.g., to identify CpG islands or repeats
- Codon usage analysis for small genomes
- Rapid identification of sequence patterns in large-scale sequence sets
- Presentation tools for publication

22.5 Molecular Visualization

It means looking at molecular models in order to explore and understand them. Molecular visualization does not necessarily involve molecular modeling, which means creating molecular models, or changing the composition or configurations of existing models. The amino acid sequence in a given protein molecule determines the structure or the conformation of a particular protein. Protein conformation in turn determines the function of that particular protein. Hence, with the conformation or structure predicted, the function of any unknown protein can also be predicted with similarity search techniques. The relationship between sequence and function is primarily concerned with understanding the 3D folding of proteins and inferring protein function from these 3D structures. Molecular visualization helps the scientists to bioengineer the protein molecules.

22.5.1 Tools for Molecular Visualization

22.5.1.1 RASMOL

The term RasMol is derived from Raster (the array of pixels on a computer screen) and molecules. This is a molecular graphics program intended to visualize proteins, nucleic acids, and small molecules for which 3D structures are available. In order to display a molecule, RasMol requires an atomic coordinate file that specifies the position of every atom in the molecule through its 3D Cartesian coordinates. RasMol accepts this coordinate file in a variety of formats, including the Protein Data Bank (PDB) format. The visualization provides the user a choice of color schemes and molecular representation. Additional features such as test labeling for selected atoms, different color schemes for different parts of the molecule, zoom, rotation, etc. have made this the most popular of all visualization tools.

22.5.1.2 Chime

Chime is a derivative of RasMol that allow visualization inside web browsers, while RasMol runs independently outside a web browser. Hence, it can be used only online, when connected to the internet. Another feature of Chime is that only certain molecules that are allowed by the company can be seen and not like RasMol where any protein molecule with atomic coordinates can be seen.

22.5.1.3 Cn3D

Cn3D is a helper application that allows us to view three-dimensional structure from NCBI's entrez retrieval service. Cn3D is a visualization tool for biomolecular structures, sequences, and sequence alignments and has powerful annotation and alignment editing features. It displays structure-structure alignments along with their structure-based sequence alignments, to emphasize what regions of a group of related proteins are most conserved in structure and sequence.

22.5.1.4 MolMol

MolMol stands for Molecule analysis and Molecule display. It includes features that are not found in RasMol and Chime. It is a molecular graphics programme for display, analysis, and manipulation of three-dimensional structures of biological macromolecules, with special emphasis on nuclear magnetic resonance (NMR).

22.6 Benefits of Bioinformatics

- Improvements in fish production and performance traits such as disease resistance to maintain profitability.
- Marker-assisted selection (MAS) in fish breeding.
- Analysis of virulence genes and vaccine development.
- Novel approaches to analyzing biological systems in the context of whole genome sequences (microarray analysis and data mining in silico).
- Understanding biological structural properties and modeling the structures.

22.7 Future Scope

- Sustainable genetic improvement of fish by identifying economically important genes and traits.
- Statistical analysis of genomic sequences of fish.
- DNA fingerprinting of fish microbial pathogens.
- Studying population genetic variation and identification of trait associated genes.
- Correspondence analysis of relative synonymous codons in fish.
- Statistical analysis of microarray data of fish.
- Modeling and forecasting of fish production.
- Designing of new molecules such as drugs, peptide vaccines, proteins, etc.

- Conservation of fish genetic resources.
- Integration of specialized tools such as geographical information systems (GIS), image analysis, and statistical and modeling software to find and track patterns, such as geographic distribution of fish populations and their diseases.
- Construction of evolutionary conservation of genes.
- Holistic ecosystem modeling to optimize sustainable fish production and management.

Genomics and Its Application in Fisheries 23

Genome is the complete genetic material of an organism (Hieter and Boguski 1997) which includes both nucleus and mitochondrial genomes. Hence the scientific discipline of mapping, sequencing, and analyzing the genomes is called genomics (Mckusick 1989). The term "genomics" was coined by Thomas Roderick in 1986. According to the central dogma of molecular biology, DNA is transcribed into RNA which is then translated into protein. When the whole set of DNA, i.e., the genome is transcribed into whole set of RNA, it is called as transcriptomics and when the whole set of RNA is translated into whole set of protein, it is called as proteomics. Hence the broad term genomics encompasses genomics, transcriptomics, and proteomics. Genome analysis is divided into structural and functional genomics (Table 23.1).

23.1 Application in Aquaculture and Fisheries

Genomics find its major application in the management of fish genetic resources (FiGR). Similarly, it has application in disease diagnosis, production of fishes with better growth rate, feed conversion efficiency, body conformation, and fillet yield through genetic improvement. Molecular markers have been developed for many farmed species which helps in genetic analysis for FiGR conservation and genetic enhancement. Linkage and physical maps have been developed to know the genes responsible for particular traits, and expressed sequence tags have been produced providing material basis for the development of microarray technology. Many farmed fish species have high fecundity that provides large full-sib and half-sib families, and this greatly facilitates quantitative trait loci (QTL) mapping.

C. J. Betsy, C. Siva, *Fisheries Biotechnology and Bioinformatics*,
https://doi.org/10.1007/978-981-99-6991-3_23

Table 23.1 Difference between structural and functional genomics

Structural genomics	Functional genomics
Studies the structure, organization, and evolution of genomes	Studies the expression and functions of the genomes
It is the basis for functional genomics	Assess the gene functions by using the information provided by structural genomics
The result is the construction of high-resolution genetic, physical, and sequence maps of an organism	It provide new insights into the behavior of biological systems
The ultimate map of an organism is its complete DNA sequence with a resolution of every single base pair	It is characterized by high-throughput methodologies combined with statistical and computational analysis of the results
Computational biology is used for data generation and management	Computational biology is used for mining the data sets for valuable biological information

23.2 Potential Genomic Technologies for Aquaculture and Fisheries

Some of the potential genomic and analytical genetic technologies for aquaculture and fisheries include DNA markers, genetic mapping, microarrays, marker-assisted selection (MAS), and sequencing. DNA markers are used for genetic linkage mapping, analysis of genetic resources, strain and species differentiation, parentage identification, and preservation of genetic diversity and conservation of genetic integrity. MAS is used for genetic enhancement; environmental improvement through enhanced productivity and environment monitoring, monitoring antibiotic resistance; diagnosis of fish diseases, vaccine development, and delivery method; stock identification and for their use as FiGR in aquaculture; conservation of FiGR, including protection of endangered species; and the development and application of transgenic fish technology. QTL analysis reveals statistically significant linkage between phenotypes and genotypes, thereby providing explanation for the genetic basis of variation in complex traits (Falconer and Mackay 1996; Lynch and Walsh 1998). It also provides an avenue for addressing some of the disease problems that threaten aquaculture since QTL mapping will help to identify genes responsible for disease resistance. Through MAS, brood stocks can be developed containing disease resistance genes and traditional hybridization can be used to allow their introgression (Senanan et al. 2004). Some of the QTL studies carried out in aquaculture are given in Table 23.2.

23.3 Major Aquaculture Genome Projects

- The NRSP-8 project for catfish, tilapia, salmon/trout, striped bass, shrimps, and oysters at US
- The Genome Research on Atlantic Salmon Project (GRASP) in Canada

Table 23.2 QTL studies of aquaculture species (modified from Liu et al. 2012)

Species	Traits
Atlantic salmon	Body weight and condition factor, growth, adaptive traits, life history, resistance against IPV and ISA
Coho salmon	Hatch timing, weight, length, and growth
Rainbow trout	Growth, upper thermal tolerance, life history, spawning time, osmoregulation capacities, development rate, whirling disease resistance, smoltification
Arctic charr	Body weight, salinity tolerance, condition factor, and age of sexual maturation
Asian seabass	Growth
European seabass	Body weight, morphometric traits, and stress response
Gilthead sea bream	Sex determination and body growth
Tilapia	Sex determination
Common carp	Muscle fiber-related QTL
Eastern oyster	Disease resistance
Pacific oyster	Growth, resistance against summer mortality
Zhikong scallop	Size-related traits
Pacific abalone	Growth-related traits
Blacklip abalone	Growth

- The SalMap project for mapping the Atlantic salmon genome in Europe
- The BASSMAP project for mapping the genome of the European sea bass (*Dicentrarchus labrax*) in Europe
- The BRIDGEMAP project for genome of sea bream (*Sparus aurata*) in Europe
- The Grass Carp Genome Project in China
- Genome projects on Japanese flounder, yellow tail, shrimps, and oysters in Japan

23.4 Fish Genome Sequencing

The whole genome sequencing was first done in Japanese pufferfish, *Takifugu rubripes* (Aparicio et al. 2002). So far genome sequencing and assembly have been done for many fish species including Asian Seabass, Atlantic cod, Atlantic salmon, Channel catfish, common carp, European seabass, grass carp, Japanese flounder, Nile tilapia, Northern snakehead, rainbow trout, tongue sole, turbot, and Pacific bluefin tuna. These sequences were conducted de novo using next-generation sequencing technologies such as 454 and Illumina and PacBio. The estimated genome sizes ranged from 544 Mb in turbot (Figueras et al. 2016) to 2.97 Gb in Atlantic salmon (Lien et al. 2016). The predicted protein-coding genes in the genome of these fish species ranged from 19,877 in Northern snakehead to 52,610 in common carp. Among these species, GC content ranged from 31.5% in channel catfish to 45.8% in Atlantic cod. The base composition is related to genomic function and species ecology (Smarda et al. 2014). In teleost, the GC content was found to be higher in seawater than freshwater fish and higher in migratory than non-migratory species (Tarallo et al. 2016).

Genes associated with specific traits can be predicted using functional genome annotation which reveals numerous molecular mechanisms underlying phenotypic traits linked to aquaculture. In grass carp, the KEGG (Kyoto Encyclopedia of Genes and Genomes) pathway analysis identified genes associated with adaptation from a carnivorous to an herbivorous diet that was likely involved in transcription activation of the mevalonate pathway and steroid biosynthesis (Wang et al. 2015). The Gene Ontology and KEGG analyses of the common carp genome suggested a significant portion of these genes associated with epithelial morphogenesis, pigmentation, and immune response, including growth factor signaling and other functional pathways (Xu et al. 2014).

References

Aparicio S, Chapman J, Stupka E, Putnam N, Chia JM, Dehal P, Brenner S (2002) Whole-genome shotgun assembly and analysis of the genome of Fugu rubripes. Science 297(5585):1301–1310. https://doi.org/10.1126/science.1072104

Falconer DS, Mackay TF (1996) Introduction to quantitative genetics, 4th edn. Longman, Harlow

Figueras A, Robledo D, Corvelo A, Hermida M, Pereiro P, Rubiolo JA, Gut M (2016) Whole genome sequencing of turbot (Scophthalmus maximus; Pleuronectiformes): a fish adapted to demersal life. DNA Res 23(3):181–192

Hieter P, Boguski M (1997) Functional genomics: it's all how you read it. Science 278:601–602

Lien S, Koop BF, Sandve SR, Miller JR, Kent MP, Nome T, Zimin A (2016) The Atlantic salmon genome provides insights into rediploidization. Nature 533(7602):200

Liu S, Zhang Y, Sun F, Jiang Y, Wang R, Li C, Zhang J, Liu Z (2012) Functional genomics research in aquaculture: principles and general approaches. In: Saroglia M, Liu Z(J) (eds) Functional genomics in aquaculture. John Wiley & Sons Inc, pp 1–40

Lynch M, Walsh BJ (1998) Genetics and analysis of quantitative traits. Sinauer Associates, Sunderland

Mckusick VA (1989) The human-genome organization- history, purposes, and membership. Genomics 5:385–387

Senanan W, Kapuscinski AR, Na-Nakorn U, Miller LM (2004) Genetic impacts of hybrid catfish farming (Clarias macrocephalus × C. gariepinus) on native catfish populations in Central Thailand. Aquaculture 235:167–184

Smarda P, Bures P, Horova L, Leitch IJ, Mucina L, Pacini E, Rotreklova O (2014) Ecological and evolutionary significance of genomic GC content diversity in monocots. Proc Natl Acad Sci U S A 111(39):E4096–E4102. https://doi.org/10.1073/pnas.1321152111

Tarallo A, Angelini C, Sanges R, Yagi M, Agnisola C, D'Onofrio G (2016) On the genome base composition of teleosts: the effect of environment and lifestyle. BMC Genomics 17:173. https://doi.org/10.1186/s12864-016-2537-1

Wang YP, Lu Y, Zhang Y, Ning ZM, Li Y, Zhao Q, Zhu ZY (2015) The draft genome of the grass carp (Ctenopharyngodon idellus) provides insights into its evolution and vegetarian adaptation. Nat Genet 47(8):625. https://doi.org/10.1038/ng.3280

Xu P, Zhang X, Wang X, Li J, Liu G, Kuang Y, Sun X (2014) Genome sequence and genetic diversity of the common carp, Cyprinus carpio. Nat Genet 46(11):1212–1219. https://doi.org/10.1038/ng.3098